T0267157

Folk Tales
of the
COSMOS

Folk Tales

of the

COSMOS

JANET DOWLING

ILLUSTRATED BY VICKY JOCHER

First published 2023

The History Press
97 St George's Place, Cheltenham,
Gloucestershire, GL50 3QB
www.thehistorypress.co.uk

© Janet Dowling, 2023
Illustrated by Vicky Jocher

The right of Janet Dowling to be identified as the Author
of this work has been asserted in accordance with the
Copyright, Designs and Patents Act 1988.

British Library Cataloguing in Publication Data.
A catalogue record for this book is available from the British Library.

ISBN 978 1 80399 417 8

Typesetting and origination by The History Press
Printed and bound in Great Britain by TJ Books Limited, Padstow, Cornwall.

Cover illustration adapted from an illustration by Mohamed Nemr, courtesy of
Shutterstock

MIX
Paper from
responsible sources
FSC® C013056

Trees for Life

CONTENTS

A Warning

Quite often books of folk tales are given to children. This is a warning to parents that these stories contain adult themes and may not be suitable for children under the age of 14.

With thanks to my life partner, Jeff Ridge, who supports me in everything I do, holds my hand when things get dark, ensures that I get plenty of oxytocin hugs and knows exactly how much green tea and chocolate frosted cake is needed.

And to my friends and colleagues at the Norman Lockyer Observatory and the audiences who listened to my retellings of the tales of the night sky. I learned so much about how the stories were working from observing how you were reacting to the told tales.

And many thanks to my dear friend Margaret E for 40 years of friendship and support in dark days and the light.

'AH! JUNE STAR'
BY S.E.K. MQHAYI

(Extract)
Summon all the nations, so that we can allot the stars:
The stars should be allocated,
You, Basotho people, take the Dogstar, the Harvest
star before winter,
And share it with the Tswana and Chopi, and other
loincloth wearers,
The Zulu will take the belt of Orion,
And share it with the Swazi, the Chopi and the Shangaan,
As well as all the other uncircumcised peoples.
You from Britain, take Venus,
And share it with the Germans and the Boers.
You whites-who-do-not-know-how-to-share-anything
Learn to share it with the Boers and the Germans.
We will hold on to the June star, we the people of Phalo,
That group of seven stars,
Is the star by which we count our years,
We count the years of being a man,
We count the years of manhood,
I can go on and on, but here I will stop.

© Translation: 2008, Antjie Krog, Ncebakazi Saliwa &
Koos Oosthuyzen

Introduction

Humans love stories – whether they are in the land, sea, sky or about people in these settings. Born with curiosity (a gift granted to us from Eve) our brains are programmed to find connections between things – a basis for survival. The easiest way to make and remember a connection is to make a story. Whether it's the Australian aboriginal who learns the landscape by endowing it with story, or the night sky full of stars to identify the ones that give you direction or tell you the time of year, there is a story. Or the cave dwellers with their paintings of the hunt – stories to be told, with the sharing of tactics that did or did not work. The observations that at certain times of the year a cluster of stars would be in alignment with the silhouette of a mountain, and soon after the rivers would flood. Or that when another cluster of stars rose above the horizon to spend time in the night sky, that foretold the season when the winds would be so great that nothing could leave the harbour. Over time these little associations became bigger and more essential to survival on a day-to-day basis.

The star clusters themselves became a calendar device where rhythms of nature were observed and facilitated the efforts of hunter-gatherers. The sun rose every day and the moon completed a monthly cycle. The Pleiades, a small, easily identifiable cluster of stars, moved above and below the horizon just before the equinoxes, becoming a seasonal marker. Hours to the dawn or sunset could be measured by the position of the moon and sun relative to the landscape. Some stars rose at the time the bison would start to migrate, or the disappearance of a cluster would lead to a period of intense cold, so preparations needed to be made for a long, hard, cold winter.

But even as you look at the night sky, it depends on where in the world you are looking from. The earliest tale-spinners were from Mesopotamia and Babylon, but they only saw the stars according to their position. To them, the night sky had part of the northern sky including the north polar star, and some of the southern hemisphere sky. From our position in England, we would have more of the northern hemisphere and perhaps a flicker of the northern lights, which the Greek had no notion of. Or in the south, the Māori would have only the stars of the southern hemisphere with no knowledge of the northern skies. Each community saw and interpreted the night sky according to their community, and what the story meant for the locality.

By the early twentieth century, the sky was full of overlapping 'recognised' constellations. In 1923, the International Astronomical Union officially recognised eighty-eight constellations. A map of the night sky was then drawn so that a star's coordinates were uniquely identified in a part of space named for the closest constellation. Thus, the night sky is a patchwork of oddly shaped 'pieces' anchored around a constellation.

The majority were the forty-eight named constellations from Ptolemy. These included many that came from the earlier civilisations of Mesopotamia and Babylon (and others), incorporated into the Greek systems, and then added to by the Greeks' own observations of the stars. The characters of the Greek myths were woven around the stars so that the sky becomes a story book. As you recount the stories, you learn the constellations and their position in the sky. As you learn the constellations, you are better able to assess their movement in the sky relative to the landscape (or specific manmade constructions like temples, standing stones) so that they are more use as a seasonal calendar or navigation tool.

I was inspired to write this book when I realised that the night sky was full of the Greek myths – that the constellations linked through story, and sometimes over two different parts of the sky. Then, as I researched them, I realised the eighty-eight were very Eurocentric in origin. I felt it was important to demonstrate that there were star stories from other cultures, but there were so many I could easily have filled ten volumes. However, many of the stories come from indigenous communities for whom the stories have a spiritual meaning. Then, balance of respect and cultural appropriation needs to be considered. Essentially, cultural appropriation is the use of sensitive cultural materials out of cultural context, and inappropriately for the purpose for which they were originally intended. In some cases, there also is the issue of monetary gain made by use of the material that does not return to the community.

There are many layers to a story. One is the 'top' story, the narrative that describes the basic story actions and consequences. Then there are further layers that relate to the description of the landscape the story takes place in, the emotional landscape between the main characters, the societal relationship, the mores and rules of behaviour, and not least the spiritual element of the relationship of the teller with the landscape. If you are part of the community, you recognise the references in the gaps between the words; but if you are not part of the community, the references go over your head and become meaningless. Many cultures do not want to share their original deep stories but may allow an outline action-only retelling – the outer edges of the top story. But changing or adding in one element of the top story through ignorance of the culture may cause offence. For example, in the Christian story of Easter, if a teller missed out the crucifixion and resurrection because 'it did not fit with their experience or beliefs', this would mean missing out a key component of the faith (i.e., that Christ died to save them from their sins and rose again).

I wanted to show that the Eurocentric naming of constellations and stars was not the only way to look at the night sky, and that because one cultural group saw a warrior and his sword, this did not automatically mean it was obvious to other cultures.

Some tradition-bearers I consulted said, 'Don't do it.' Some other tradition-bearers said, 'A lot of it is out there already – do it but be respectful in what you do.'

At the same time Equity, the performing arts and entertainment union (www.equity.org.uk), was drafting its Guidelines on Storytelling and Cultural Appropriation. I consulted with several members of the group doing the work. The key principles of the guidance are available online, and this is my short version:

- Respect the culture.
- Research to understand the culture including symbols, language, images, or landscapes.
- Learn from the culture.

As part of this process, it is important to understand the transition stories have been through from a native teller to the resources that you are working with. Ursula McConnel, in *Myths of the Munkan* (1957, based on field work in 1927 and 1934), says:

> The stories are consistent with their environment and the social order which binds this environment to the storytellers. This blending of natural and social factors, so strange to us, cannot be understood in terms of external values … To appreciate the inner logic of the stories one must be familiar with their background … It must be realised that these stories are for oral transmission and the more active parts for dramatization. It is not easy to do justice to their style in the verbose form of written English, nor to convey the telling pauses, the pregnant silences. The impressive reiteration and the innuendos of the speaker's voice, who, as he relates the story experiences and visualises the acts and scenes embodied in his short dramatic sentences … I have adhered as closely as possible to their way of telling.

Some storytellers will tell you the work that goes into telling a story starts with identifying the bones of an action line.

It is then rehydrated with description of both characters and landscape, the emotional journey of the characters, the physical gestures that add to the story, the tone of the voice and the intentions of the storyteller. All affect choice of words and the delivery of the story.

However, in a written collecting situation you have the teller, the person who translates the story to the collector, the transcription into written words initially in note form, then in expanded form, and finally edited for publication. From first to last the emotion, intentionality and physicality are unlikely to be conveyed in the manner the storyteller intended. Furthermore, the listener may relate a part of the story to their personal experience and then carry that image forward when they retell the story themselves. An example of this is Devil's Tower in Wyoming. Originally known as 'Bear Lodge' by native peoples, in the nineteenth century it was first described in English as 'Bear Lodge Tower'. It may have been transformed phonetically into 'Bad God's Tower' and thus into 'Devil's Tower', which has a completely different meaning, has Christian connotations, and is offensive to the local community. At the time of writing (2023) there are attempts to get the name changed to reflect the original, which has more meaning to the local people.

There is also the influence of European colonisation and missionary work spreading their faith through stories. Once a story has been told and then retold, the links with the original storyteller and the context of their telling get lost. In retelling the story, the teller draws on aspects of language, concepts and landscape that they are familiar with. The listener/collector has no idea if this is a story from within this culture, if it has been influenced in the near or distant past, or just simply made up by a skilled storyteller in front of the listener using existing motifs and landscapes. Several nineteenth-century collectors, across the world, cited stories of the seven stars (aka Pleiades) where maidens run from a great hunter in different culture as evidence of a worldwide ancient origin myth – whereas it only takes one person retelling the Greek myth of Orion and the Pleiades for the story to be embedded in the culture.

For example, A.W. Reed (1999, p.298) retells the Australian Aboriginal story where first man and first woman are told by the creator to use everything they need in the garden, but not to touch anything of the tree that has the bee swarm and honey. The woman collects some wood and realises that she has picked up some fallen branches of the forbidden tree. Nothing happens to her. She is then emboldened to try the honey, at which point a giant bat, placed by the creator to guard the tree, emerges. Death has now entered the world 'by the evil the woman has done'. This has some elements of the Adam and Eve story, as well as the tale of the golden apples placed by the Greek goddess Hera in a garden and protected by the dragon. Is this an origin story of the forbidden fruit in paradise – or a story influenced by a more recent local retelling of either the Biblical or Greek version? As a storyteller, I might let my personal perspective influence the stories I tell depending on my intention of the set. As a collector and reteller of traditional stories, I would be as neutral as possible – avoiding any influence of my own culture, values and prejudices.

In making this selection and retelling them, I have given myself the following ground rule.

I would try to determine the potential source of the story and how far it is from the original spoken source. If the story source was close to the root story, then my retelling would reflect the same language without any emotion, morals or values, landscapes or spiritual reference that was not in the original story. If it was a source that had clearly undertaken many retellings, and I could not find a root source, then I felt free to introduce emotion, personal values and intentionality.

Stories from the Greek myths were based on local variations and old myths from the Egyptians and Mesopotamians. Over the past 2,000 years they have been retold and reinterpreted, and I felt that gave me licence to retell the stories in my own way. The versions of the Greek myths of the night sky are the ones from my research of the many variants of the myths that I prepared for my storytelling performances entitled 'Lust and Revenge: Tales of Passion in the Night Sky'. There are, of course, other Greek myths that are not represented in the night sky. Some stories from European and Chinese roots also had many variants, with none claiming to be the original. I chose to draw on the variants and craft my own.

Finally – this is a collection of folktales and myths of the night sky. It is not a comprehensive anthology of all the many and various tales – that would take several volumes. In the end, I could only skim the top layer of stories from each continent and give a taste of the wide variety of tales told and the richness of imagery beyond the Greek Eurocentric vision. However, I have included 'Notes on the Stories', which gives details of the sources I have used, and other books that I consulted to understand the context, which I hope readers will use as a springboard to do their own research.

If there are any errors or places where my writing is insensitive, I apologise now, and request that you contact me at JanetTells@ gmail.com to advise me.

1

THE GREEK MYTHS
OF THE STARS

The myths of the Greek gods have been handed down and became the role models by which subsequent generations and civilisations learned how to behave and how to be a leader.

Zeus was King of the Greek Gods, and with his two brothers, Hades and Poseidon, and sisters Hera, Demeter and Hestia, they were known as the Olympians. They had overthrown their father Cronus and his brothers and sisters, known collectively as the Titans. Thereafter, it was Zeus and the Olympians who held the praise and the loyalty of the people on earth. They forged their allegiances within the family – marrying each other, brother to sister – and having children born immortal and thus gods themselves.

However, some of the male gods, during the span of their immortal lives, sought their pleasures with more earthly women. Their Olympian wives either turned a blind eye or took out their vengeance on the earthly mothers and children. Never their husbands. A god does not contradict a god.

A child born of god and human had a chance of being born immortal, or not. There was no way of predicting which was likely. At birth, immortality was evident. The chosen ones would be feted, and when old enough to leave their mothers, they were taken to Olympus, home of the gods. But those who were mortal were left to make their own way in the world. Some of them became heroes, and some were souls blighted by their sense of loss, knowing that their birthright could have been so very different.

The Greeks populated their night skies with characters from their myths and legends. Chief among them are the stories of the mortal sons of gods, who raised themselves up from being abandoned sons to achieve reputations that made them immortal in memory.

THE GREEK MYTHS AND THE CONSTELLATIONS THAT APPEAR IN THE MYTHS

Myth	Constellations within the myth
Perseus: hero of the people	Medusa, Perseus, Ladon (the Dragon), Pegasus (the flying horse), Cassiopeia (the vain queen), Cepheus (her husband), Andromeda (their daughter and wife to be of Perseus), Cetus (the great sea monster)
Orion the Hunter and Ophiuchus the Healer	Corvus (the crow), Ophiuchus, Centaurus, Orion, Pleiades, Canis Major, Canis Minor, Scorpion, Taurus
Callisto	Great Bear, Little Bear
Hercules: the twelve labours	Hercules, Milky Way, Centaurus, Leo (the first labour), Hydra (water snake, second labour), Cancer (crab, second labour), Sagitta (the arrow), Corvus (the crow, aka Stymphalian birds, fifth labour), Taurus (the seventh labour), Draco (eleventh labour)
Jason and the Argonauts	Centaurus (the centaur), Aries (the ram), Argo (the ship; including Puppis the stern, Carina the keel, and Vela the sail), Castor and Pollux aka Gemini (the twins), Eridanus (the river)

PERSEUS: HERO OF THE PEOPLE

Medusa, Perseus, Ladon (the dragon), Pegasus (the flying horse),
Cassiopeia (the vain queen), Cepheus (her husband),
Andromeda (their daughter and wife-to-be of Perseus),
Cetus (the great sea monster).

Medusa's Story

Along the coast of Greece there were temples dedicated to the
Goddess Athena served by virgin handmaidens. There was one
young woman, Medusa, who wanted to serve Athena. She was
taught the rituals, the offerings, and her duties when she was to
stand to pray to Athena. Down by the shore there was a certain olive
that grew there. With the sea salt, it looked and tasted so beautiful.
The Goddess Athena particularly enjoyed them as an offering.

Sometimes, when the wind is up, waves are frothy and white.
People called them white horses. As Medusa gathered the olives,
she could see a real white horse emerging out of the water onto
the shore, stand and then shake itself. She walked up to it and
put her hand out to stroke its mane. The horse turned to her
and started to nuzzle her neck. She found herself leaning into the
warmth of the body and the muscles of the body, holding it tight.
There were thoughts and stirrings in her that felt unfamiliar. On
an impulse, she pulled herself up onto the horse, which raced up
the beach. She rode until the sky darkened, and it was time for
her to start the evening rituals at the temple. She slipped off the
horse and began to walk back, taking the olives. As she came up
to the steps of the temple, the horse followed behind her.

'No, sorry, you can't come in here,' she laughed. She put her
arms around its neck and held it tight, enjoying the moment with
her eyes closed, breathing in the scent of the horse.

Then the smell changed. She opened her eyes and was no longer
holding a horse. It was a man. She could see his aquamarine eyes

and knew in that moment this was Poseidon in his human form, the God of the Sea, brother of Zeus. He stepped towards her and said, 'You have ridden me. Now it's my turn.'

She stepped back into the temple for sanctuary. But Poseidon followed her and there, on the altar of Athena, he took his pleasure from her. She tried to fight him off but there was nothing she could do. In the moment of his joy, he stood up, laughed, then walked away. Medusa cried out to the Goddess Athena, 'Please help me.'

Up in the heavens, Athena looked down. A heavy musk hung in the air, and she knew what had happened. 'My altar has been defiled. This is no longer a place of sanctuary. You have destroyed this.'

Medusa tried to explain: 'I did not know it was Poseidon, I thought it was a horse.'

'He is my uncle,' responded Athena. 'It must be your fault!' And with that, she cursed Medusa.

Her nails went yellow, turning into claws. Her limbs began to grow scales. She put her hands to her face. Scales there too. Her beautiful hair was now a seething mess of hissing snakes.

'No man or woman will ever look at you without out being turned to stone. You will be hated and hunted for all time. *Get out of my sight!*'

Medusa staggered out of the temple. There were people coming to lay offerings. They saw her, and with no time to even to gasp, they had turned into stone. She did not know where to go. She fed herself in the night from the plants that she could scavenge in the woods. Always running. Behind her was a trail of stone statues. Young men, determined to prove their bravery, declared, 'I will go and slay this Medusa!' Many never found her, many never returned.

Eventually, Medusa came to a shore. On the horizon she could see an island. She took a small boat and let it float across. As she came to that island and got out, two creatures came towards her with scaly skins, claws, and snakes for hair. 'We are the Gorgons,' they said. 'We are immortal daughters of the gods. We have heard of you, Medusa. Come with us, be our sister, even though you are human and mortal. Share with us our pain.'

Medusa stayed, perpetually disturbed by the growing number of stone statues of young warriors on the shore determined to seek her death.

In the night sky, you can see the head of Medusa as part of the constellation of Perseus.

Perseus' Story

There were a king and queen who had a daughter, Danae. However, kings of this time wanted sons to follow them to create their own dynasty. It was the fashion to ask for advice from the Oracle of Delphi – priests who would 'consult' with the gods and give a reply. The king went to the Oracle and asked, 'Please tell me, when will my son be born?'

The Oracle replied, 'You will have no sons; you will not have any other daughters. Your grandson will slay you.' The king was horrified; no sons, no other daughters. The only way he could have a grandson was if this small child, Danae, would grow up and have a son. He did not know what to do. He certainly did not want to kill his daughter. His solution was to place her in an underground chamber, forged out of bronze and lit by candles, hidden out of sight. He told his wife he had sent the child away for safety. Handmaidens were appointed, sworn to secrecy, who were tasked with the care and upbringing of the child. His only thought was: 'She must never, ever meet a man.' The years passed, and Danae grew up. The king did not know what else to do to prevent a grandson being born. He visited her and was the only man she ever saw. Danae had only vague memories of what it was like in the world before she came into this underground bronze chamber.

Zeus, King of the Gods, always liked a dalliance with challenge. From his throne in the heavens, he could see over the whole world. When he saw Danae in a bronze chamber underground, hidden from sight, he knew he had to meet this one. He would show that he could not be kept out of anywhere. In keeping with the bronze chamber and the filtering lights, he turned himself into a golden shower and seeped through the earth, down into the chamber, and flowed all over the girl. As it touched her skin, Danae laughed and cried. The handmaidens cowered in fear. They did not know what was happening. Danae embraced it, and just for a moment she felt like there was someone hugging her back. Then it was gone.

'Did you see that?' she asked the handmaidens.

'What was that?' The handmaidens shook. 'We saw nothing.'

In time, her belly bulged. When the king saw his daughter, he was furious. He quizzed the handmaidens until one confessed. 'It was a golden shower; we think it was Zeus.'

Danae's father shook with rage. He feared for his own life but decided he would not personally kill his own daughter and grandson. Instead, he would leave it to the will of the gods. He ordered Danae to be locked up in a box, taken out to the sea, and then the box thrown overboard. If she drowned, it was the will of the gods – not his.

Zeus had a choice. Whilst he felt he could not intervene himself, he would help things along. The box floated until it reached the nets of a fisherman called Dictys, who pulled it up out of the sea into his boat. When he opened it, he was very surprised to see a woman suckling a newborn child in it. He took them to shore, fed and cared for them. Danae told him that she was the daughter of a king, and this was her son Perseus, a child of a god. Dictys wasn't interested in any of that. He just wanted to know that she and the child were safe. His brother was King Polydectes, but Dictys was content to be a fisherman. He was kind, and he was gentle. Danae was happy to stay with him and to learn how to wash, cook and sew.

In his palace, Polydectes heard that his brother had 'acquired' a woman and a child. He came down to see what was going on. 'Well brother, what have you here?'

As soon as Polydectes learned that Danae was a princess in her own right, he wanted her as his wife and queen. He said, 'Come to my palace. Marry me. It is a fitting match and much better for you than staying with my useless fisherman brother.'

Danae was frightened and said, 'No. I cannot marry you until my son is a man. I must look after him.'

Polydectes laughed. 'Well, when your son is grown, then I will come calling.'

Over the years, Perseus grew up. He learned all the ways of the sea, as well as how to fight and how to be strong. His mother told him all the myths and legends of the kings of Greece, not just the gods. With these stories he grew and learned how to work with men and women, and how to face trials and tribulations.

Eventually, Polydectes came and said, 'Your son is old enough. Marry me.'

Perseus responded, 'My mother does not need to marry you. And for as long as I'm here with her, she will not.'

'We shall see,' said Polydectes. He returned to his palace, and next day there came an announcement that he had decided to marry a princess from overseas. She would arrive soon, and everyone was to give her a gift of a stallion. Polydectes invited his brother Dictys and Danae to the wedding.

Perseus was expected to attend. He sent a message, and said, 'I do not have the money to buy a stallion. But give me a task and I will do it. I will bring you anything.'

Polydectes laughed because all of this was a ruse. There was no princess from overseas. He was just looking for a way to get rid of Perseus. He was very pleased to declare, 'Bring me the head of the Medusa!' Perseus was stunned but he agreed. 'Good,' said Polydectes, 'for when you do not return, we will know either you have failed or you are a coward, too frightened to return.'

Perseus went to the temple of Athena, and cried, 'Athena, please help me in this. How can I take the head of Medusa without being turned to stone?'

Athena appeared with Hermes, messenger to the gods, by her side. 'Perseus,' she said, 'I've heard your cry. Medusa has been such a thorn in my side. It is good to hear that you wish to end her time on this earth. Alas, too many young men have fallen to her sight and are in stone. So, I will help you. What you must do is avoid looking at her head, the snakes, and her eyes. Use this polished shield, which is like a mirror. This sword is made of the strongest substance that we have. It will cut through her neck with one blow. But you must be careful,' she said. 'You need to get swiftly in and out, making no mistake. Medusa lives with the Gorgons. And they fly fast, too.'

Hermes interrupted. 'For this, I'm going to lend you my second-best pair of flying sandals. Here they are.' The little wings on the sandals fluttered. 'And this bag,' he said, 'will be very helpful to put the head in!'

'Beware,' said Athena. 'The Gorgons have very good eyesight and will see you at great distance. You will also need the Helmet of Hades to make you invisible. It is with the Hesperides who guard Hera's tree of golden apples, with Ladon the dragon. However, you will not find the Hesperides unless you can find Atlas, who holds up the world. They are his daughters.'

'Where will I find Atlas?' asked Perseus. Hermes smiled and whispered in his ear.

Finally, Perseus felt he was ready. He thanked both Athena and Hermes for their help. 'Is there anything else?' he asked.

Athena hesitated, then said, 'I have a message for Medusa. Give it to her before she dies.' Athena whispered in the other ear.

Perseus put on the winged boots and flew into the sky. Getting his bearings from the sun, the landscape and the stars that were just coming out, he went in search of Atlas.

In one of those twists of reality, Perseus found Atlas holding the Earth on his shoulders. This was Atlas's punishment from Zeus for being one of the twelve Titans, the previous race of gods that Zeus, with his brothers and sisters, the Olympians, had overthrown. As Perseus landed, Atlas looked round and grimaced.

'Hello,' said Atlas. 'Have you come to hold this for a little while?' indicating the world on his shoulders.

Perseus said, 'No.'

'I've got a scratch, an itch,' responded Atlas. 'Please can you…?'

'Certainly,' said Perseus. He came around the giant, climbed up, and just scratched.

'Oh, that's so good. I've lived with that scratch for a thousand years. Are you sure you won't hold this for just a moment?'

'No,' said Perseus. 'The Hesperides, your daughters. I need to ask them for the Helmet of Hades. I need it. I'm going to seek the head of Medusa.'

'Medusa, poor girl.' Atlas shook his head. 'Perhaps that's the best thing for her.'

Atlas told Perseus where his daughters were, and he flew up into the air.

The Hesperides are the goddesses of the evening sunlight. You can see them each night as the sun goes down. They sang such beautiful songs and guarded the tree of the golden apples with the dragon Ladon underneath.

You can see Ladon in the night sky by the pole star.

The Hesperides kept a watch on the sky and saw something come flying towards them. They got ready with their arrows and their spears. Perseus called out, 'Your father sent me.' With that they laid down their weapons. 'I have come for the Helmet of Hades, the cloak of darkness that will make me invisible. I seek the Medusa to cut off her head.'

One of the Hesperides looked at the others and said, 'Athena did badly with that one, didn't she?' Another said, 'Poor Medusa. All that time in that body, all that fear and that rage. We must help her.' The third said, 'Yes, take the Helmet of Hades.' Perseus became invisible as soon as he put the helmet on. He rose in the air, but they had very good ears. They listened to him as he went, and they sang him a song to be glorious and compassionate.

It took some searching, but eventually Perseus found the island of the Gorgons. Wearing the Helmet of Hades, he hovered above the beach where there were stone statues of young men in different states of fight or flight. On the shoreline, there were three creatures, with scaly skins, yellow claws, and writhing hair. Two of them were very large, and one was just about human size. The smaller one remained on the beach while the others went back into a cave. This was Medusa, her head tilted back, scouring the sky.

'I can hear you,' she said. 'I cannot see you. But I can hear you. Do not think you will get away from me. You will not get away from my gaze!'

Perseus hovered. His shield polished as a mirror in one hand, the sword in the other. The bag from Hermes on his belt, tied for a quick release. He was ready. But he had a message to deliver.

'Why can't I see you?' screamed Medusa. 'I know you are here. I can hear you and smell you! Look at me. Join these other men!'

Perseus saw her reflection in the shield. He could feel the fear that was instilled in so many men before their death, even though he was protected. 'I have a message for you!' he called. 'From Athena!'

Medusa fell to her knees. 'She has not forgotten me.'

'Athena said to tell you, "Me too! Forgive me!"'

With that, Medusa raised her arms, pulled back her head of snakes. 'I forgive her. Take me, release me! I cannot live like this.'

With one swipe Perseus brought down the sword. Medusa's head swung up into the air, and then fell into the bag held out by Perseus, guided by the mirrored shield. He closed the bag, tied it to his belt, and lifted the sword. He was ready to fly away when the other two Gorgons come out of the cave. 'What is this? What has happened to our sister? She was the one who was mortal. She was the one who would die. But not yet!' They saw the drips of blood from a sword they could not see. They could hear a noise. They unfurled their wings as they launched themselves into the sky. As fast as Hermes' winged sandals travelled, the Gorgons were catching up with Perseus.

The blood from the head of Medusa soaked through the bag and fell into the crashing white horses on the seashore. As the blood hit the sea there was an explosion, and out of it emerged a flying white horse.

He rose towards Perseus, who climbed onto its back, held on to the mane, and together they went up into the sky. The Gorgons

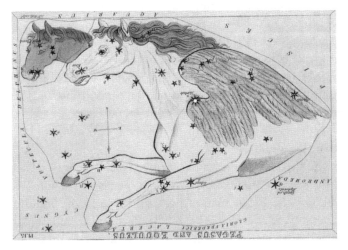

This is Pegasus.

reached out to just touch the edge of the tail of Pegasus, but with the extra incentive the two flew up into the sky with the head of Medusa safely in a bag.

King Cepheus decided that he wanted to marry the most beautiful woman in the world. He had a competition, and after a lot of deliberation, he chose one. 'Yes, you are beautiful. You may come and be my wife.' Her name was Cassiopeia. The more beautiful she was told she was, the more beautiful she believed she was. The couple had a daughter, Andromeda.

As Andromeda grew up, everybody said how they were like sisters. 'My daugh-

Cassiopeia.

ter and I are so beautiful,' crowed Cassiopeia. 'We must be even more beautiful than the Nereids. They're supposed to be the most beautiful people in the whole world. But I'm sure my daughter and I are much better than that. Aren't we, my dear?'

Andromeda was more down to earth. 'No, mother, don't boast,' she would say.

The Nereids were the handmaidens to Poseidon, the God King of the Sea. They heard these boasts and complained to him. 'We will not have any humans saying that they are more beautiful than us. We're not going to be your handmaidens anymore unless you do something about this.' Poseidon wasn't used to this kind of solidarity between the Nereids.

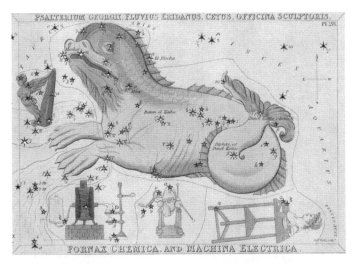

Cetus is sometimes shown as a whale, and can be found in the night sky.

'I know what I'll do,' he said. "I'll send Cetus, the sea monster.'

Cetus was very big. Sea waters flooded before him over all the land, making it salty and impossible for the crops to grow. He did this again and again. The land crops were ruined. The fish in the sea were scared away. There was going to be a famine. The water that humans and animals tried to drink was foul and rank. The people were so afraid they were going to die, they pleaded with King Cepheus to do something.

He did the only thing he could – he sent someone to the Oracle at Delphi for advice. The Oracle replied, 'Sacrifice your daughter, Andromeda.'

Both Cepheus and Cassiopeia were horrified and tried to sacrifice another young woman. But Cetus, the monster, ignored her, and continued to wreak havoc. Andromeda was not pleased to hear what her parents had done. 'Mother,' she said, 'it was your boasting that got us into this, I have to get us out of it.'

Andromeda allowed herself to be chained to the rock outside the city to wait for the sea monster. She had a plan – she was

Andromeda.

probably going to die, but she wasn't going to go down without a fight. While she stood waiting for Cetus to appear, she could see something in the distant sky. It looked like a man on a flying horse. It was Perseus on Pegasus.

'Hello,' she said, as Perseus hovered before her.

'Are you in a spot of bother?' he asked.

'No,' she said, 'I'm perfectly fine, thank you very much. I've got this.'

'Are you sure you don't need some help?' replied Perseus.

'I'm fine. I'm just going to be eaten by a monster, I'm the sacrifice. But don't worry, because I'm going to hold on to these chains, raise myself up, and give him a good kicking so that he'll give up and go home. Then I'll have saved everybody.'

Perseus said, 'I have a suggestion.'

'What can you do? This sea monster's very big.'

Perseus replied, 'I've got the head of Medusa.'

Andromeda gasped. 'Really?'

He said, 'Why don't you do your plan, and then when you kick him so hard that he rises out of the water, then I'll do my plan. But when you kick that monster, don't forget to close your eyes. You don't want to look at Medusa's head.'

Andromeda waited. She could see Cetus in the distance, rising to the surface of the water, making its way towards her. It opened his mouth and she saw the rows of teeth. If this didn't work, she would be shredded. As it came towards her, she braced herself with those chains and kicked the lower jaw of the sea monster. It leapt up out of sea into the air with eyes wide open, to see Perseus with Medusa's head. In that moment the sea monster turned to stone and fell back into the waters.

'That worked,' laughed Perseus. 'We're a good team, aren't we?'

She looked at him and she looked at the horse. 'Maybe,' she said. 'Who is asking?'

They both mounted Pegasus' back and flew back to the palace. Cepheus and Cassiopeia were already organising the celebration. To be brief, there was a fabulous feast to mark the end of the tyranny of the sea monster, and a celebration of the marriage of Andromeda and Perseus. But partway through the celebrations, Perseus remembered something. He whispered to Andromeda. She looked around at the wedding party ... and nodded. 'Go now, no one will miss you.'

Perseus went and found Pegasus. Together they flew into the sky, and soon they could see Atlas holding the up the Earth.

'You came back!' groaned Atlas.

'Yes,' replied Perseus, 'I had to return the Helmet. I just called in to see you.'

'Could you just hold this a minute? Give an old man some rest?' Atlas asked the age-old question about the planet on his shoulders. Perseus said nothing. There were tears in his eyes. Atlas asked, 'Is that Medusa's head in that bag on your belt?' Perseus nodded. 'I'll be back,' he said. Atlas sighed. 'Tell the girls I'd like to say goodbye.'

Perseus flew to the Hesperides and gave them back the Helmet of Hades. They spoke together for a while, then all travelled back to see Atlas. 'Are you sure?' asked Perseus. Atlas nodded. His daughters embraced him in turn, turned their backs to him with their eyes closed, and sang a song that soared from the bottom of the sea to the height of the heavens, singing their father's praises. At a nod from Atlas, Perseus closed his eyes, opened his bag, held up the head of Medusa, and then replaced it. Atlas was now a god of stone, holding up the world but no longer in constant pain and discomfort.

Perseus returned to the wedding after Pegasus dropped him off outside the feasting hall. Pegasus' work was finished, and now he rode the sky to Olympus as the immortal son of a god. Perseus slipped back into the celebrations. No one had noticed he had left.

'Is it done?' asked Andromeda. Perseus nodded.

The wedding feast was glorious and went on for three days. In time, there was a baby born called Perez, after Perseus. Perez eventually grew up to be the King of Persia. The newly wedded couple lived with the king and queen, with Cassiopeia even learning to curb her tongue.

Perseus had a dream about his mother, so together with Andromeda they took a ship and eventually came to the island where he'd left her. 'Stay there,' he said to the crew and Andromeda. With the head of Medusa strapped to his belt, he strode across the beach. The fisherman's hut was empty. He went to the palace. Outside the gates, he could see Dictys sitting with a beggar's bowl.

'What's happened?' Perseus asked.

Dictys was shocked. 'We thought you were dead. Polydectes has taken your mother and plans to marry her tomorrow. But today, he's feasting.'

Perseus smiled. 'Oh, well I have something for him!' and he strode into the banqueting hall. There was Polydectes and his cronies. 'Here I am,' declared Perseus, 'I have returned!'

Polydectes turned round and said, 'So, you're not dead? But I don't see the head of Medusa about you. You are a coward who never achieved anything. Now your mother will be my wife.' Polydectes and his cronies laughed.

Perseus put his hand into his bag, closed his eyes, and lifted Medusa's head up. Laughter turned to silence in an instant. Perseus replaced the head in the bag, and then opened his eyes. All the company had turned to stone.

Perseus found Danae locked away in an underground chamber. She wept many tears when he found her and brought her out into the light. The first to greet her was Dictys, who was now king.

He asked her to marry him, and she said, 'Yes! From the moment you opened the box I loved you, but for the shame I could say nothing.' And they were wed.

Perseus introduced everyone to Andromeda, and they all had a big celebration. They had to build another banqueting hall first, because the first one was full of statues.

Now Perseus' thoughts turned to his grandfather. He took a ship, and sailed to the island from whence he came. As he got off the ship, he found everyone was talking about 'the games'. The old King had no heir, and the victor of 'the games' would be his successor.

Perseus took part and excelled in every single sport. Javelin, archery, sword fighting and wrestling. Everyone wanted to know who this stranger was! The last game was discus. When he threw it, into the air, it was clear it was going to be the furthest of all. However, the fates decreed that a seagull flew in front of it, diverted it, and it hit the King on his forehead. He fell to the ground, dead. Perseus was taken and brought before the elders and advisors.

He said, 'I am the victor of these games. It was not my fault that the seagull diverted the discus.'

The elders could understand that. 'This must be the gods' way of telling us that this man should be our king!' they said. 'But, who are you?'

'I am Perseus, the grandson of this King. It was foretold that I would slay him. He abandoned my mother to the seas with me in her belly. I am returned now.' Perseus paused. 'I came here to meet him. I did not intend to cause him harm, but it is as the Gods will it.'

All was revealed and so Perseus was made King. And in time he had three kingdoms to rule over with his wife Andromeda. But the one thing he knew was that whatever the Oracle tells you, it doesn't always work out that way. So, he always took the counsel of his wife – because that, at least, he could rely on.

ORION THE HUNTER AND
OPHIUCHUS THE HEALER

Corvus (the crow), Ophiuchus, Centaurus, Orion, Pleiades,
Canis Major, Canis Minor, Scorpion, Taurus.

Ophiuchus' story

Apollo, son of Zeus and Leto, God of the fine arts, medicine,
music, poetry and eloquence, was rather enamoured of Coronis,
the mortal daughter of the King of the Lapiths. With his words,
poetry and eloquence he seduced her, made her his plaything,
and she became pregnant with his child. Apollo was pleased at
the thought of a son to ride with him in his chariot carrying the
sun as it passes over the daytime sky. He sent his spy, Corvus, the
white crow, to watch over her and report back her movements.

The words of seduction are quickly forgotten when overtaken
by true love, and Coronis' heart was given to a mortal man whom
she loved with much joy. Corvus the crow observed all this and
reported back to Apollo. The god was outraged. He stomped his
feet and behaved like a petulant child. 'Why didn't you pluck her
eyes out!' he shouted at the crow. Apollo was so angry he turned
the crow the black, and that is why crows are black today.

Corvus the crow is a constellation in the night sky.

Apollo, God of the Sun, obsessive, turned to the one person
he could trust. His sister, Artemis, Goddess of the Moon but also
Goddess of the Hunt. He complained of Coronis' 'treachery'. In
love and loyalty to her brother, Artemis let loose her arrow to pierce
the heart of Coronis, who died with the babe in her belly. As was
the custom of the day, her body was laid out on a funeral pyre.

From his chariot in the sky, Apollo watched as the flames were
lit. In a moment of contrition, he turned to Hermes, Messenger to
the Gods. 'Save my son,' he said. 'He may be mortal, but he does

have the blood of a god.' Hermes, in his winged boots, swept down, cut the baby from her body, and watched as the flames grew higher to absorb it.

Apollo had no need of a child who was not mortal and thus not destined to join the Gods on Olympus. So, Hermes took the child to Chiron, the centaur – half man, half horse – who cared for many mortal sons of the gods, as well as the sons of kings of the earth.

Chiron understood the pain and pleasure at being the son of a god. He himself was born of a liaison between a nymph and Cronus – Zeus' father and king of the old order of Gods, the Titans. To avoid being detected, Cronos had turned himself into a horse, and thus the child Chiron was conceived: a half man, half horse. Unlike many of the sons of gods, Chiron was immortal, but under the new order he kept his peace and taught young men to be the best they could be.

Chiron the centaur, one of two centaurs in the sky.

Chiron looked at the babe in Hermes' hands. 'This child is fresh born and has never tasted his mother's milk. Far too young to take part here. I have no time to care for his needs, and there are no wet nurses here. Take him away.' Hours old and already rejected twice, the baby opened his eyes and looked at Chiron. He moved his hand towards the old centaur, who sighed and grimaced as the tiny fingers clasped his thumb.

'Just as seductive and compelling as his father, I see,' said Chiron. 'I'll take him if you will provide the means to feed and nourish him.' He reached out for the baby, who then nestled on the centaur's shoulder.

Hermes smiled. He pulled from his pouch a small goatskin. 'This is full of the milk that Zeus, his grandfather, was nourished with as a baby. It will replenish once a day and should suffice.' Chiron nodded. Cradling the mewling child in one arm, he opened the goatskin pouch to allow a few milk drops to fall on the lips of the child. The crying stopped as the little tongue found the nourishment he craved.

The baby was named Ophiuchus. He was small for his age and was kept in a basket, and taken by Chiron to every class that he taught. Although much younger than the other boys, he was attentive and learned well. He was above average in most of the classes – combat, sword fighting, music and art, even politics and strategy – but his great love came to be medicine. It intrigued him how most of the failings of the human body were rapidly soothed by local plants. He wanted to be a healer, restoring people to good health. When it was time to leave the school, Chiron gave him the goatskin bag. 'This is yours now. It is rich with nourishment and will serve you and your patients well.'

Out in the world, Ophiuchus soon established a reputation for healing people with the milk and herbs he collected. At first, he wandered on foot from town to town, but soon he responded directly to requests for help.

King Minos had a young son, Glaucus, about eight years old. He was playing a game with his friends when he fell into a large standing jar of honey. By the time help was summoned, Glaucus had stopped breathing. Nothing could revive him. In despair, King Minos summoned Ophiuchus and asked him to restore his son, but that was beyond Ophiuchus' skills and abilities. Instead, he offered to stand vigil by the boy's bed. As he knelt in the light of the flickering candles, he prayed to Zeus to give him guidance.

Zeus, in his throne at Olympus, heard the prayer. He looked down, and recognised Ophiuchus as a mortal grandson. He contemplated for a moment, waved his hand, then sat back and watched.

In the darkness of the night, Ophiuchus was startled when, attracted by the sweetness of the honey, a snake slithered across the ground towards the bed. Outraged, Ophiuchus killed the snake with one blow from his staff then returned to his vigil.

After a time, he saw another snake slither across the floor. This snake held a leaf in its mouth and placed it on the dead body of the first snake. To his surprise, Ophiuchus saw the dead snake twitch and come to life. The two snakes disappeared into the undergrowth. Intrigued, Ophiuchus retrieved the leaf. It was not one that he recognised, so he spent the rest of the night looking among the bushes to find the original plant.

As the sun came up, he saw the morning dew glistening on a small plant hidden in the roots of a common bush. It was the same. He took another leaf and placed it on Glaucus' lips. As the king entered the room, Glaucus opened his eyes, called for some water, then greeted his father. King Minos was astonished and offered Ophiuchus anything. All Ophiuchus wanted was a root and specimen of the plant.

Already well known for his healing knowledge, with this plant amongst his medicines he could now bring back the newly dead. Wherever he went, he administered to both poor and rich without distinction.

Hades, brother to Zeus, God of the Underworld, seethed. Not only had he been deprived of Ophiuchus' soul at his birth, but now because of Ophiuchus' gifts Hades was seeing fewer souls in his kingdom. He was afraid that his power and authority would diminish. He was not pleased that this was due to the actions of Zeus. So, he watched and waited as Ophiuchus made his way through the world.

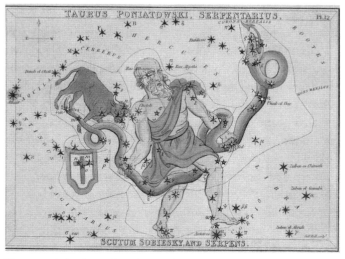

You can see Ophiuchus in the night sky as he wrestles with the serpent.

Orion the Hunter's Story

Poseidon, brother to Zeus and God of the Sea, had a liaison with a sea nymph. From this was born Orion. A giant of a man, but merely mortal. Poseidon had no interest in his son, but Orion swore to make his father recognise him and be allowed to live on Olympus – a privilege given only to immortal gods.

As he grew from youth to man, Orion trained to be the strongest and the fastest. He could shoot an arrow and hit a target a mile away. His sense of smell developed so that he could track an animal without even seeing it, letting go his arrow and being certain of the shot. He would walk on sea beds, even across the surface of the water, such were his powers of breath control and lightness of foot. But no matter what he did, his father Poseidon took no notice of him.

He fell in lust with Merope, the sole heir and daughter of King Oenopion of Chios. King Oenopion was himself a son of the god Dionysus by Ariadne – so a god's blood ran through Merope's veins. In Orion's mind it made perfect sense to marry the sole heir to a king who was the son of a god himself. In time, the throne would come to her husband – Orion! Surely his father would have to pay attention to him if he eventually became a king by marriage.

King Oenopion felt under pressure and agreed to the marriage on condition that Orion killed all the wild beasts on his island. He thought it an impossible task. Orion stood in the centre of the island, cocked his ear, and systematically let loose his bow. One by one he killed the wild beasts until none were left. Now Oenopion was worried about the implications for himself if he had such a powerful son-in-law. So, he reneged on the deal. Orion was so enraged that he no longer saw Merope as his lover but took revenge on her for her father's action. He raped her. For the custom of the day, it meant that even though she was blameless, it was her virginity that was more important. It would be difficult for her to find someone to marry her. For the king, it meant that his dynasty was at an end. He sought his revenge by giving Orion so much drink that he became oblivious. Whilst unconscious, the king's men blinded Orion and placed him on a boat sailing far from the island.

From Olympus, Hephaestus, God of Blacksmiths, looked down. A son of Zeus, he saw his cousin's predicament. Having some sympathy for Orion's struggle, he sent his servant to guide Orion towards the east to catch the healing rays of the morning sun. With him he took two hunting dogs. They were to guard Orion from coming to any harm while he was blind. When his sight was restored, they became his hunting dogs. At the same time, Hephaestus had sympathy for Meriope and Oenopion. He built them an iron citadel where Orion could not find them to perpetuate the cycle of revenge.

Orion the hunter with his club, killing one of the wild beasts.

Orion became lost and disaffected. Nothing seemed to work for him. He was frustrated in his every move, and still his father, Poseidon, did not recognise him. He wandered aimlessly across the world, with his dogs, looking for opportunities to make his mark.

On his travels, he came across the Pleiades: the daughters of Atlas. He was a last remnant of the old order of Gods, the Titans, and an uncle of Zeus. His punishment from Zeus, for fighting against him, was to hold up the sky. With no father to protect them, Orion saw the daughters as easy prey and pursued them, hoping to make one, or more, of them his wife. But the sisters would not be separated, and they fled together. Angry and frustrated, Orion chased them, trying to persuade any one of them to join with him.

Atlas, confined to holding the world, saw what was happening. He called out to Zeus: 'I take my punishment, but do not let my daughters suffer for my actions. They flee from Orion who will not stop his relentless pursuit of them. Please help them.'

Zeus, who in his time had dallied with one or two of the sisters, listened. With a flutter of his fingers, the sisters turned into doves and flew into the sky. Orion tried to use his strength to leap and reach them, but they raced up and away beyond this earth into the night sky. They became the stars known as the Pleiades.

The Pleiades can be seen for half the year and are an important marker for the seasons.

Dissatisfied with his life, and still not recognised as the son of a god, Orion resolved to find Artemis, daughter of Zeus. As well as Goddess of the Moon, she was Goddess of the Hunt. Orion, in his befuddled state, reasoned that if he could excel against a Goddess, his cousin, surely he could be raised up as a God himself?

It took many years, but finally he found her. She was very amused by him, and when he proposed to have a competition to see who could hunt and kill most beasts in the hour before midday, she laughed and agreed. She was surprised that he matched her beast for beast. When the midday sun beat down on them, she turned to Orion and said, 'Enough now! You are a far better hunter than I gave you credit for, and much better than any other man or god. The sun is high, let us stop now and we will call this a draw.'

Orion did not hear her. He was focused on what he was doing. He did not see the sun high in the sky. He did not hear Artemis telling him to stop. All he knew was in that moment he was in his prime, matching the Goddess, showing his strength. As he killed the beasts, he named them as a sacrifice to his father Poseidon.

The Scorpion.

With Orion unresponsive to her calls, Artemis called out to Gaia, the old Goddess of the Earth, for help.

Gaia was horrified that this braggart was killing beasts in her domain without any remorse or purpose. It was a waste. She resolved to rid the earth of this man. She created a scorpion from the sands of the desert, that clambered out of the rocks at the feet of Orion.

Leaving nothing to chance, she directed a great bull to charge towards Orion. With a smile on his lips, Orion decided to use this as another show of his prowess. He would stand his ground, show his nerve, never falter, and at the last possible moment he would let loose his arrow. With all his concentration on the bull in front of him, he did not notice the scorpion by his foot. Nor the sting that pieced his skin not once but three times. It was when the poison started surging through his leg muscles that he became aware something had happened. He crashed to the ground, fitting as the poison reached his heart. The last thing he saw was the scorpion. The bull stopped in its tracks, trampled over the bow, and went back to grazing.

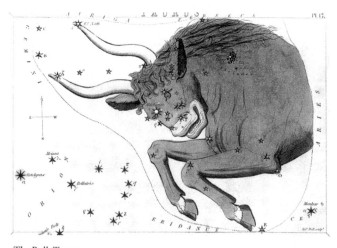

The Bull, Taurus.

Orion's breathing stopped. As his eyes glazed in death, Artemis wept for him. Truly he had been a match for her. She had not intended for Gaia to kill him, but each God to their own.

Artemis summoned Hermes and asked for his help for Orion. Hermes knew just the person, and brought Ophiuchus to administer his life-giving herbs. Ophiuchus used his staff to first kill the scorpion and then knelt by Orion, opening his bag to prepare the life-restoring herb.

Canis Major, one of the two dogs. The other is Canis Minor.

At this point Hades, God of the Underworld, was even more unhappy. He called out to Zeus that this was unfair. Yet another soul was being prevented from coming to his domain.

Then Zeus spoke to Ophiuchus. 'Wait, I have not decided whether he will live or die. Your plants work by my pleasure, and I have not decided.'

Artemis argued that Orion should be given another chance – after all, he was the son of a God, and he was a gloriously skilled hunter. But Zeus looked at the path of destruction that Orion had spread across the world. The number of times that he himself had to intervene.

'No, I will not allow him to live yet again. I will recognise that he is the son of a God, and he has shown his great skill, but his arrogance has had too many consequences. I will place him in the night sky, along with his two dogs, to be a warning to those who would boast.'

Gaia, in her turn, was not happy. Zeus had placed Orion in the sky close to the Pleiades. To protect her granddaughters, she placed the bull, Taurus, between the Pleiades and Orion, and then placed the scorpion in the sky below Orion's feet ready to nip him if he got out of control again.

Hades was furious that another soul had escaped his kingdom of the underworld, but decided to use this to his advantage. He whispered in Zeus' ear that Ophiuchus had claimed the power of life and death for himself, and that while Zeus was all powerful, he could not keep an eye on the use of Ophiuchus' herbs all the time. So why let a mere human be as powerful as a god?

Zeus was seduced by his brother's words. His rage was fuelled at the thought that a human could have the powers of a god, and in anger unleashed a thunder bolt that killed Ophiuchus. But in that moment, Zeus knew that this was not a good act.

To make recompense, Zeus placed Ophiuchus between Orion and the scorpion, so that Ophiuchus could perpetually guard Orion. At the same time, Zeus placed the stars of a snake in his hands – the one that had shown him the leaf. A reminder of both the healing powers but also that some things are still the gift of the gods.

CALLISTO

Great Bear, Little Bear.

Callisto was a daughter of King Lycaon. She chose to become a follower of Artemis, Goddess of the Hunt and of Childbirth. Although human, Callisto grew very skilled with hunting and running, and was a favourite companion of Artemis when she was in the field. Like Artemis, Callisto swore to be chaste and never marry. Zeus, King of the Gods, saw vows of chastity as a personal challenge.

One day, Callisto was resting in the earthly woods, enjoying the flowers and the gentle wind, when Artemis came and sat by her side. This was nothing unusual and often happened. The attention from Artemis became more intimate and demanding. Callisto was not sure what was required from her. Then, suddenly, the nature of it all changed, and to her horror Callisto realised this was not Artemis but Zeus in disguise. She fought him off, but to no avail, and when he left her she sobbed, afraid and unable to tell anyone.

Over the next few weeks Callisto carried out her duties, but she was always tired and sluggish. She was no longer the first person to go hunting with Artemis, now she held back. When Artemis proposed that they go to the seashore and swim in the water, all the other companions agreed and clapped their hands with joy. Callisto watched as they stripped off their clothing and ran into the water. She held back again, but some of the companions pulled her towards the water, taking her clothes off. As she stood naked, her full belly was seen and the implications were obvious.

'You have been with a man!' cried Artemis. 'You have broken your vow of chastity. I cannot have you contaminating my companions with your loose behaviour – you must leave here now!'

Artemis picked up some stones that were just under the water and started throwing them at Callisto. The companions copied her, and Callisto ran from the water, wet and bloodied, trying to put on her clothes. She wandered around for several days until she collapsed in front of an old woman's hut. The old woman took her in, and seeing her state set a bed for her. Callisto told her what had happened, and then started to give birth. The old woman prayed to Hera, wife of Zeus, for support and guidance. But when Hera looked down, she realised that Zeus had betrayed her with this earthly woman, and that the baby was his son. With one swipe of her hand, Callisto was turned into a bear. She blundered out of the hut, fearful of the harm she might do her baby son. Zeus could do nothing to counter his wife's actions.

The old woman took the child back to Callisto's father, who took the boy on as his own son as he had no other heir. The boy was called Arcas. His grandfather taught him to hunt, fight, wrestle – all attributes a king should have.

Arcas was as good a hunter as his mother. He heard that there was a bear that had been roaming the local woods, and resolved he would prove his manhood by killing it. He went into the woods, armed with his bow and arrow. The bear was Callisto, who had kept close all those years to know how her son was. In a clearing, Callisto and Arcas came face to face. She knew who he was, but he had no inkling of who she was. He raised his bow and arrow, waiting for the moment to take his best shot.

In the meantime, Zeus had cast a curious eye, and realised what was happening. Whilst he could not counter Hera's action, he could create a new one of his own. So, with swipe of his hand, he turned Arcas into a bear, and then placed both bears in the night sky with Callisto as the Great Bear, and Arcas as the Minor Bear.

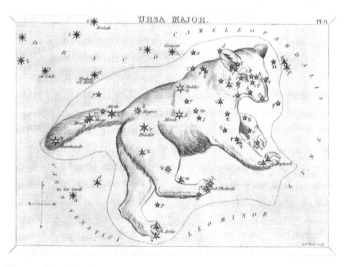

The great bear in the sky.

HERCULES: THE TWELVE LABOURS

Constellations: Hercules, Milky Way, Centaurus,
Leo (the first labour), Hydra (water snake, second labour),
Cancer (crab, second labour), Sagitta (the arrow),
Corvus (crow aka Stymphalian birds, fifth labour),
Taurus (the seventh labour), Draco (the eleventh labour).

Zeus was in lust again. Alcmena was the granddaughter of Perseus, the great hero. She married Amphitryon, a general in the army, who was returning home after a long period at war. Alcmena was looking forward to his arrival, and when he appeared a day earlier than she expected, she was very pleased and took him to her bed. What she did not know was that this was Zeus, in disguise, pretending to be her husband. There was great confusion the next day when her real husband arrived – and she took him to bed too! She was soon pregnant with twins: one from Zeus and one from Amphitryon.

Zeus declared a descendant of Perseus would be born, who would become King of Mycenae – one of the largest Greek kingdoms. He intended that this should be the child of Alcmena.

Hera, Zeus' wife, was furious. She was the Goddess of the Hearth and Marriage, and yet Zeus made a mockery of her. She was also Goddess of Childbirth and knew that another earthly woman had conceived a child who was also a descendant of Perseus. Hera arranged that this second child would be born at seven months – before the son of Zeus. This first child born was Eurystheus, who went on to become King of Mycenae. The younger, second child born, son of Zeus, was Hercules.

When Zeus' son was born, he was originally named Alkides for his earthly grandfather. His mother, fearful of Hera, abandoned him in the woods. The goddess Athena was watching and took the baby to Hera, asking for her help to feed the foundling. Hera, flattered and not realising the child was Hercules, took him to her breast. Her milk would give him strength and power, but when he accidently bit her, she pushed him away. Milk from her breast was spilt across the sky and became the Milky Way. Horrified, Hera gave the child back to Athena, telling her to care for him herself. Athena returned Hercules to his mother with the promise that she, Athena, would look out for her son.

Having deprived Hercules of the right to the kingdom, Hera continued to find ways to bring down Zeus' son. Hercules and his twin, Iphicles, were in their cradle. Hera sent two poisonous snakes to kill them both. Hercules crushed the life out of them. People nodded their heads – this was evidence that he would be a great hero.

The twins were sent to Chiron, the centaur, to learn archery, fencing, chariot racing and wrestling, as well as reading, writing and singing. Hercules exceled at sports, winning every competition he entered, but he had little patience. When his lyre teacher tapped his arm when he played a wrong note, Hercules lashed out and accidently killed him. Hercules was remorseful. For retribution, his stepfather, Amphitryon, commanded him to guard his cattle at Mount Cithlaeron. It was on this mountain that Hera sent two young women to tempt Hercules. One was Pleasure, who offered him a life of idle luxury. Hera hoped that Hercules would choose this, having neglected to mention that excessive pleasure leads to a downward spiral into vice. The other was Virtue, who offered him a life of toil and suffering, and eventual happiness through serving others. To Hera's surprise, Hercules chose Virtue.

A lion stalked and killed the cattle that Hercules guarded. He in turn hunted and killed it, presenting Amphitryon with the skin of the beast. His reputation grew, and the King of Thebes was pleased when his daughter, Megara, agreed to marry Hercules. At first their life was good, and they had three sons together.

Hera watched. It was a complete affront to her that Hercules had a good life. She was in such a frenzied state about Hercules that she made Hercules go mad. When he wrestled with his sons, he was overtaken by a belief that they were his enemies, out to kill

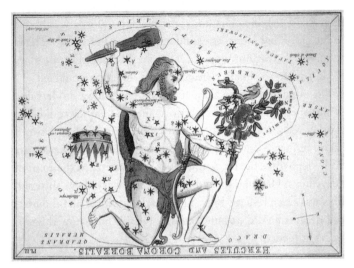

Hercules.

him. He fought them with all his strength. Megara screamed for him to stop, but he saw her as an enemy too. He killed them all.

When the haze of madness lifted, Hercules looked at what he had done and was filled with horror and fear that he was so out of control. In grief and remorse, he wanted to kill himself to atone for their deaths. Friends told him that he had not been of right mind, and they forgave him, but he could not forgive himself.

The Oracle at Delphi was a place people went to seek advice, believing that words of the gods came from the mouth of the priests. When Hercules asked the Oracle, he was told that he could redeem himself by subjecting himself to the rule of his cousin, Eurystheus (the other baby who was now King of Mycenae) to complete ten tasks. To Hercules this made sense, remembering his music teacher and the task he had to undertake. He thought he would get the peace of mind he sought. For Hera, this would allow her to influence Eurystheus to create opportunities where Hercules could lose his life and have her work done for her. It was the Oracle that told Hercules to change his name from Alkides to Hercules, to celebrate 'Hera's glory'. Would this be the glory that Hercules would gather after a lifetime of Hera's attempts to bring him down? Or would this indeed be the glory of Hera when she finally defeated him? Only time would tell.

The other gods were aghast at Hera's venom towards Hercules. Whilst they would not directly go against Hera, they could help in other ways. Athena brought him a robe to protect him, a sword from Hermes, a bow and arrows from Apollo, and a breastplate from Hephaistos the Blacksmith. When he arrived at the court of Eurystheus, Hercules looked magnificent. Eurystheus shivered. This was not a responsibility he had asked for. He had thoughts and dreams he did not understand. He did not know this was Hera's influence.

The First Labour: The Nemean Lion

The Nemean lion was a child of the Goddess of the Moon, Selene. Spurned by her, it roamed on the earth, creating mayhem and havoc in the land of Nemea, killing and devouring not only cattle, sheep and horses, but also men, women and children. No sword, knife or arrow could penetrate its skin. Anyone who tried to master this menace died before the day was out.

Eurystheus thought this would be the ideal task for Hercules to lose his life. Hercules accepted the challenge, and with his bow and arrows, sword and club, began the search for the Nemean lion. Its lair was in the mountains, with two entrances. Hercules blocked off one with huge boulders, then faced the lion in the cave. It came towards Hercules, its jaws open and watering. Its roar echoed through the cave. Hercules swung his club above his head and gave a great blow to the side of its head. For a second the lion was stunned and distracted. Hercules leaped forward, wrestled the lion to the ground, and with all his strength he squeezed his hands around its neck. He struggled as the lion thrashed back and forth, until finally the light went from its eyes, and it died.

Hercules took one of the claws of the lion and stripped the skin from its back attached to the head. He wore it as he strode across the country-side, back to Eurystheus. His cousin quaked as he watched Hercules approach. He'd always felt like an imposter, as if someone else should be king. With Hercules in his court and the dreams he had from Hera, he felt very threatened. He found his favourite hiding place as a child – a bronze jar, that

The Nemean lion is shown the stars as Leo, one of the signs of the Zodiac.

was just big enough to fit him as a man. Its cool surface against his forehead calmed him down as he wondered what the next task should be.

The Second Labour: The Hydra of Lerna

The Hydra of Lerna was a serpentine beast that lived near the city of Lerna. It had seven heads like snakes that hissed and spat, with grotesque teeth that tore the flesh of its victims. Its breath was deadly. Every time one of it heads was cut off, another grew in its place.

Eurystheus thought this was the beast that would finish Hercules off. He gave the order, and Hercules gathered his belongings to make the journey, taking Iolaus, his nephew, with him. After a long journey, they found the cave where the Hydra lived. Iolaus set a fire, and Hercules let the flames light up his

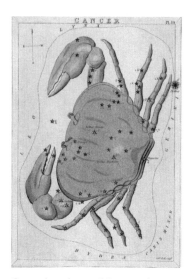

Cancer the crab, one of the signs of the Zodiac.

arrows, shooting them into the cave to entice the hydra out. It emerged spitting and angry at being disturbed. Holding his breath and wielding his sword, Hercules got close and struck off one of its heads. It regrew instantly. Hercules turned away to gasp for air. Hercules approached again, wielded his sword, and cut off a head. This time Iolaus spang forward with a fiery branch and cauterised the stump before it regrew. It worked!

Hera, watching from above, was furious! All she needed was for Hercules to be distracted for a moment, and then he would be forced to breathe in the toxic fumes. She sent a crab to nip at Hercules' ankles – surely that would work. But Hercules was focused on the Hydra. The crab continued to pinch his ankles. Hercules leapt up into the air, spun around, took another gasp of air, and as he landed brought down his sword across the back of the crab, splitting it in two. Then he swooped his sword up again to cut another head from underneath, which Iolaus promptly cauterised.

The last head was more difficult. It was immortal – even cut from the body it would be alive and a threat. Iolaus dug a deep hole. As Hercules cut off the last head, he caught it on the tip of his sword and tossed it into the hole, where Iolaus buried it. The body of the hydra was slumped on the ground. Hercules took all his arrows and dipped them into the blood of the hydra, the source of the toxicity in its breath. Very carefully, Hercules replaced them in their quiver. They would be useful in the future.

Sagitta is the constellation that is the arrow. It appears in many stories, but is the most potent when used by Hercules.

Eurystheus banged his head against the wall. Why couldn't Hercules die! Eurystheus found his cool bronze jar quite appealing.

The Third Labour: The Erymanthian Boar

The Erymanthian Boar was a mighty beast that had killed Adonis, son of the goddess Aphrodite. Eurystheus rationalised that as it had already killed one son of a god, then maybe it would kill two.

Hercules went in hunt of the boar, found it on the lower slopes of Mount Erymanthus, and then chased it up onto the higher slopes. Snow slowed down the boar and it had barely enough energy to put up a fight when Hercules chained it.

When Hercules presented the boar to Eurystheus, it squirmed on the floor of the palace. 'Get it out,' Eurystheus screamed. The boar was released and swam no one knows where. Eurystheus went and found his bronze jar.

The Fourth Labour: The Hind of Ceryneia

Artemis, Goddess of Hunting, had sought five of the best reindeer in the world to draw her chariot. Their antlers shone like gold, most befitting for a goddess. But one escaped, and even Artemis, swift of foot, could not catch it.

'Bring me the Hind of Ceryneia!' declared Eurystheus. If the goddess Artemis could not catch her hind, then Hercules could not possibly do so. It took Hercules over a year to finally track it down. Finally, it collapsed exhausted. Hercules bound its feet together, slung it over his shoulder, and made his way back to the palace. He had not got very far when a shadow fell over him. He looked up and there was Artemis, standing in his way.

'That is my hind!' she said. 'Thief! For that impudence, I will slay you!'

Hercules put the hind on the ground and sat down on a nearby boulder. He gestured for her to sit by him. Artemis had been expecting a fight and was surprised. When Hercules told her about the labours, she realised that he was her half-brother through Zeus, her father.

'Brother,' she said, 'take the hind – complete the task. Then bring it back to me so that it may run free, I enjoy the chase.' Hercules was happy to agree.

Eurystheus was terrified when he heard the story. 'Take it back! Take it back now!' He already had one goddess haunting his dreams, he did not want the curse of another!

The Fifth Labour: The Stymphalian Birds

On a lake near the town of Stymphalos gathered a flock of birds that were the size of the ibis, with sharp, piecing beaks, feathers of metal that the birds could aim and shoot as arrows, and claws made of brass that would tear flesh. They terrorised the town and killed many people. No one could get close to them and live, nor any armour give protection. Eurystheus thought this was just the task for Hercules and sent him there to get rid of them.

At first Hercules could not work out how to deal with these birds, but Athena came to him with a rattle made by the God of Blacksmiths, Hephaistos. It was so loud that it scared birds into the sky. Hercules climbed to the top of the nearby mountain, where he could see the birds, and released the rattle. The noise it made resounded for miles. Bewildered and confused by the sound, the birds flew aimlessly. Hercules took his time with his poisoned arrows to pick them off. The remaining flock flew far away. Hercules retrieved the arrows, then took ten of the dead birds back to the palace. Eurystheus shook his head – this had been too easy.

Some people say that Corvus the crow also represents the Stymphalian birds.

The Sixth Labour: The Augean Stables

Augeas was King of Elis. He had more cattle, sheep and horses than anyone else, with enormous stables to keep them in. He reasoned that if his animals did not get sick, then he wouldn't have to keep them in clean conditions. So, for thirty years there was no cleaning out of the muck in the stables.

Eurystheus thought about this. Clearing out thirty years' worth of muck would certainly keep Hercules away from his own palace for a good long time.

Hercules used his strength to achieve many of the deeds, but he also had a brain – and rather than shovelling muck, he thought about it. He went up the nearest mountain and surveyed the scene. He went to King Augeas and offered to clean the stables in just one day for a portion of the cattle. King Augeas was amazed. He did not believe it could be done, but as the stench and flies of the stables were overwhelming the kingdom, so he could see the value in agreeing.

Hercules took down the walls at each end of the stables, so it looked like a long tunnel. Then he dug a trench from one of these holes, through the forest to the river flowing below, and another to the upstream part of the river. As Hercules dug the last bit of the trench, the water broke through, and the river was diverted through the stables, washing all the debris downstream. By sundown the kingdom smelled very sweet. The task was finished, but King Augeas declined to pay him, saying there was not a written contract.

When Hercules arrived back at Eurystheus' palace, the king was furious. 'You have cheated,' said Eurystheus. 'You were hired to do this job; it does not count.'

Hercules was not surprised that his cousin had found out about the contract. 'But I didn't get paid!' he retorted.

With Hera whispering in his ear, Eurystheus responded: 'When you were labouring, you were doing it with the expectation of reward. This was not done for redemption. You will need to do another task.' Hercules reluctantly accepted this.

The Seventh Labour: The Cretan Bull

Poseidon was the brother of Zeus who had dominion over the sea. King Minos of Crete was a devoted follower and Poseidon presented him with a magnificent bull. He intended that Minos should sacrifice the bull to himself. However, Minos had other ideas: he kept the bull for himself, and sacrificed another instead. Poseidon was outraged, and sent the bull mad. The beast became out of control, and to get it out of his palace Minos ordered it to be released into the countryside, where it terrorised people, stalked and killed them.

Word soon came to Eurystheus. All the tasks so far had been within his local region, the Pelopnnese. Perhaps the answer was to send Hercules far away to the south.

'Bring me the Cretan bull!' he said.

Hercules journeyed to the south and asked King Minos if he may capture the bull. 'Yes!' said King Minos.

Chasing beasts was second nature to Hercules. All he had to do was pursue it until it dropped from exhaustion. Relatively easy for Hercules, who tied the feet together and slung it over his shoulders. He carried it back to the palace and laid it at the feet of Eurystheus, who rapidly retreated to the safety of his bronze jar.

The bull was initially kept in a field outside the city. Not surprisingly, it broke out and made its way to the hill of Marathon, where much later Theseus would encounter it.

The bull is in the night sky as Taurus.

The Eighth Labour: The Mares of Diomedes

Diomedes was the son of Ares, God of War. He took pleasure in inviting travellers to feast with him, wrestling them to their death, and then watching his four horses eat their flesh. The mares were so wild and strong that they could only be tethered by chains to a bronze manger. No traveller was safe in passing through that place. Eurystheus thought this would deal nicely with his problem. It was far north, and he gambled that, being the son of the God of War, Diomedes would have the strength to beat Hercules. There was a real chance that he might be relieved of this cursed mission.

Hercules travelled to the palace of Diomedes. Delighted to see him, Diomedes ordered a feast. They ate well, and then came the wrestling. Hercules had been confident in his wrestling skills, but he had underestimated Diomedes, who was more than a match for him. They wrestled all through the night until the dawn, when Hercules got him into a hold, cracked his neck, and Diomedes was dead. Hercules carried the body down to the stables and threw it into the bronze manger. The four mares devoured him in seconds, and then were suddenly calm. It just took the flesh of their master to lose their blood lust. Hercules broke their chains, fastened them to a chariot, and rode back to Eurystheus.

The Ninth Labour: The Girdle of the Amazons

The Amazons were women warriors who lived on the south shore of the Black Sea. The men of their tribe lived on another island, meeting with the women once a year. Amazons cut off their right breast so that they could throw their deadly javelins and shoot their arrows, but kept the left breast to suckle their children. They were led by Queen Hippolyta. She proudly wore a belt given to her by Ares, God of War, who admired the bravery of the Amazons.

Eurystheus had a teenage daughter who wanted Hippolyta's belt. With Hera nagging him in his dreams on the one hand, and his daughter's pleading on the other, he decided to send Hercules eastwards to bring back the belt. 'How ironic it would be,' he thought, 'if it was my daughter's task that was to be the end of Hercules.'

Hercules took a crew of men with him, expecting a battle. But he was the only man who dared step foot on the island. Hippolyta, in her turn, was quite enamoured of him, and if circumstances had been different, he might have stayed longer. He explained about the task, and to keep him one night longer she agreed to give him the belt.

Hera, watching from Mount Olympus, was incensed. She disguised herself as an Amazon and ran to the place where Hippolyta was feasting with Hercules. She called out that the men in the ship were attacking the Amazons to kidnap the queen. The Amazons immediately went on the defensive, their arrows drawn, their javelins poised. Acts of love suddenly changed to acts of war. There was a dark battle between the men and the Amazons. Hercules loosed one of his poisoned arrows, which caught Hippolyta, and she died in his arms.

Hercules did not understand how things had changed so quickly, but he suspected Hera's involvement. Nothing was to be gained by riling against her, so he took the belt and left.

Eurystheus received the belt and noted the blood on it. He washed it off before he gave it to his daughter, who squealed with delight, wore it for a few days, and then discarded it.

It was the first task that touched Hercules in an emotional way.

The Tenth Labour: The Cattle of Geryon

In the far west, on the island of Erythera, were a herd of red cattle said to be the finest in the world. They belonged to a man called Geryon, who claimed to be the strongest in the world, having three heads and six arms within the body of one man. The cattle were in the care of a man called Eurytion, who had a fierce dog with two heads.

Eurystheus really wanted the red cattle. Hercules made the long journey westwards. He ran out of land at the place we know as Gibraltar. He built tall two pillars of stone, one on either side of the straits, in order to guide him back. These became one of the Seven Wonders of the World: the Pillars of Hercules. He dedicated them to Apollo, the Sun God. Apollo was very flattered, and when Hercules asked to borrow his golden globe to use as a ship to sail across the seas, Apollo was pleased to oblige. Hercules set up sail using his lion skin that he always wore. The winds caught it, buffeted him this way and that, then a straight course across the water to Erythia. Bounding from the boat, and wrapping the lion skin around him, he made his way to the top of the nearest mountain. There he could look over all the islands. He soon saw the two-headed dog and the cattle grazing in the deep forest. It was surprisingly easy to overcome first the two-headed dog and then its master, Eurytion.

Hercules herded the cattle towards the beach, intending to load them onto the globe ship. But on the beach stood Geryon. Each of his arms possessing a sword, his three heads shouting, 'Come get me you mere mortal. I am the son of a god; you shall not beat me. I shall tear you to pieces and feed you to the fish.'

Hercules was feeling tired. It had been a long day, and the cattle were demanding all his attention. He really did not want a fight. He got out his bow and his poisoned arrows and loosed them off. One. Two. Three. Geryon looked startled, dropped his swords, brought his hands to his three throats, and fell to the ground dead.

With the red cattle loaded on the ship, Hercules returned to Spain, guided by the two pillars he had built. He returned the globe ship to Apollo. The next day, Hercules started herding the cattle back from Spain back to the palace of Eurystheus. He had completed ten tasks, and he began to think about his life and what he wanted to do.

But Hera was not pleased. 'He has not completed the ten labours!' she declared. 'His cousin Iolaus helped him with one. He was hired to do another. They do not count. He must do two more!'

Eurystheus spent the day in his bronze jar, working out how to tell Hercules.

The Eleventh Task: The Apples of the Hesperides

While Zeus was King of all the Gods and Hera was his wife, it was not always this way. In the beginning, Gaia, Earth Mother, had created all the world. It was from her that came her sons and daughters who were the first Gods: the Titans. But the rule of those Gods was cruel, and eventually Zeus overcame and banished them. He married Hera, his sister, as was the custom of the day. In recognition of the new order of Gods and celebration of the marriage, Gaia gave Hera the gift of three golden apples, which held the power of immortality. Such was their power, Hera knew she had to keep them safe.

At the end of the world, in a secret location, Hera set up her own garden. The apples were planted in the ground and grew into a tree bearing even more golden apples. She asked the Hesperides, daughters of the God Atlas, to tend the garden for her but not to eat them. An apple fell from the tree and rolled towards the sisters. One picked it up, brushed it off, and began to eat it. It was delicious! Another of the sisters reached up to an apple on the tree and tested it to see if it would fall into her hand. It did. Hera was furious. The sisters apologised, promised it would not happen again, and they would tend the tree as they had been tasked. But Hera was now suspicious and commanded the dragon Ladon to guard the apples on the tree, as much from the Hesperides as from opportunistic travellers.

Eurytheus was getting desperate. He had set Hercules tasks within his own realm, then set tasks to the far north, south, west

and east corners of the known world. Each time Hercules succeeded Eurystheus quaked and was fearful of what Hera might
do to him. Then it came to him. The golden apples in the Garden
of Hesperides at the end of the world. If Hercules broached the
garden and took some apples, surely Hera would have a legiti-
mate reason to take her revenge directly rather than use himself,
Eurytheus, as a conduit for her rage.

Thus, for the eleventh task, Hercules was to bring back three
golden apples.

After many adventures, Hercules came to the gardens. There
he found Atlas, supporting the sky on his shoulders. Hercules had
been told that only Atlas could get past the dragon, that it was not
his fate to kill it. Atlas had been holding the sky up for so long he
was very tired. When he saw Hercules, he invited him to hold the
sky for him. Hercules offered to do so if Atlas would get him three
apples from the tree. Atlas was surprised – his daughters cared for
that tree, and he didn't want to get them in trouble with Hera.
However, his back ached so much. He entered into a binding agreement with Hercules to bring back the apples. Slowly, carefully, the
world was passed to Hercules, who found it challenging even with
his strength. There were moments when Hercules thought that
Atlas would not return. Eventually, Atlas came back with the three
apples. He had been taking advantage of the time to have tea with
his daughters, who he had not seen for a long time.

Having returned, the promise was fulfilled. Atlas suggested
that he himself would take the apples back to Eurystheus to fulfil
the task. Hercules' muscles were seizing up.

'That seems fair,' he said. 'Why don't you do that. I just need
to find something to cushion the sky on my back. Could you
hold this for me?' Atlas was so excited at the thought of getting
away from the task, he willingly took the earth back.

Hercules picked up the golden apples. 'I am sorry my friend,'
he spoke. 'I cannot do as you ask – my destiny lies elsewhere.' He
turned and began the long walk back down to the plains of the
earth. Behind him he thought he could hear sobbing.

> Ladon is represented by Draco in the night sky. Some people say
> that the Hesperides are represented by the stars in Ursa Major,
> and the tree of the golden apples by the stars in Ursa Minor. Just
> as the north star is the pivot around the which the stars seem to
> spin, the Hesperides are also the element around which many of
> the mighty Greek myths spin too.

Hercules presented Eurystheus with the apples. Even though he was acting as the agent for Hera in her revenge on Hercules, Eurystheus realised he was in dangerous territory. He gave them back to Hercules, with the instructions to return them to the gardens. Hercules did not want to risk meeting Atlas again, so he called upon Athena, who arranged the return.

The Twelfth Labour: Cerberus, the Guardian of the Gates of the Underworld

Desperate to be rid of Hercules, Eurystheus set the ultimate task: to journey to the Underworld and bring back the three-headed dog, Cerberus. No one ever returned from the Underworld alive. Hercules called upon Hermes and Athena for help one last time. They agreed to guide him to the Underworld, but first he had to learn about the rituals that he would have to observe to safeguard himself in that realm. He studied the secret Eleusinian Mysteries that would allow him to enter safely, and then with Hermes and Athena at his side he found the entrance to the Underworld. Alone, he took the path to the palace of Hades, God of the Underworld. On the way he encountered many monsters and ghosts of heroes. Finally, at the Court of Hades, he was granted permission to take Cerberus, but only if he could overpower the monster by his own strength – not with any weapon or help from the Gods. It was a harder battle than Hercules had anticipated, but finally he emerged into the daylight with Cerberus tied up on his back.

Eurystheus was furious; he ran to find his safety jar, and told Hercules to get rid of Cerberus. Hercules made the long journey back, releasing Cerberus at the entrance to the Underworld, where it ran back and resumed his position as guardian with a vengeance.

Having completed the twelve labours, Hercules was released from his punishment. He went on to have many adventures and even another family. Hera was outraged, and continued to find ways to bring him down.

It is the story of the twelve labours you can see in the night sky.

Jason and the Argonauts

Centaurus (the centaur), Aries (the ram), Argo (the ship, including Puppis the stern, Carina the keel, and Vela the sail), Castor and Pollux (the twins, Gemini), Eridanus (the river).

A small boy listened in the darkness as his parents argued. He shivered at what he heard. His mother saw him, snapped at her husband, King Aeson of Ioclos, and then took the boy to his bed. She slipped in beside him, held him tight until he slept. Then she prayed.

'Hera, Goddess of the Family. Watch over my son and protect him. Keep him from harm, until he can return to his family and in time take his rightful place on the throne.'

On Olympus, Hera heard the prayer. Looked down to the earth, and up to the future. She saw the paths that were open to the boy, where he would need a nudge here and there to be great and glorious. She saw an opportunity for revenge on an old adversary, who defiled one of her temples. She nodded and took note.

The next day the mother took the boy to his father. For his own safety the boy was sent away to be schooled by Chiron, the greatest scholar, healer and sportsman of the age, and a centaur. A man to his waist, then the body and legs of a horse.

In the night sky you can see two centaurs. The poet Ovid tells us that
Chiron is represented as Centaurus that can be seen in the southern
sky. The other one is Sagittarius.

With great haste the boy, Jason, was taken away by a rider,
who travelled night and day to deliver the boy to the caves of
Chiron. The boy did not know that within a few days his father
was deposed by his uncle, Pelias.

Chiron looked at the scrap of boy. With his sense of prophecy,
he knew that in time Jason would be a great king, but first he
needed to develop the skills to become one. As for many heroes
before and after him, Chiron taught riding, shooting, hunting,
reading, writing, star mapping, the harp and healing. During
the day, Jason challenged himself to be the best he could, then at
night he would listen to Chiron tell stories. Jason's favourite tale
was of the Golden Fleece.

*A long time ago, in the kingdom of Boetia, Nephele, the cloud
maiden, fell in love with King Athamas. They married and had
two children: a son, Phryxus, and a daughter, Helle. But as much
as she enjoyed living on earth, the realm of the Gods called to her,
and she returned to the sky. In time the king fell in love again and
married a woman, Ino. All was well, until she also had two children by the king. She became jealous of Phryxus and Helle, and
did not want them to inherit the kingdom over her own children,
so she poisoned the wheat seed. When the harvest failed, and the
people were starving, her husband was persuaded to seek advice
from the great Oracle at Delphi. But Ino bribed the messenger, and
he returned with the 'message' that 'the curse could only be lifted by
the blood sacrifice of Phryxus and Helle'.*

*Convinced that there was no other way, King Athamas prepared
the sacrifice. Nephele, the cloud maiden, watched over her children from above. She was horrified and she sent a flying ram with
a fleece that was golden. As the sword was raised, it flew down,
swooped up both children, and flew them away. When the golden
ram started to descend, Helle, the girl, lost her grip, tumbled down*

into the water below, and drowned. The spread of water was then called the Hellespont after her.

The golden ram delivered Phryxus, the boy, to King Aeetes, who welcomed him and brought him up. To thank Zeus for the safe delivery of Phryxus, the golden ram was sacrificed. In turn, its golden fleece was given to King Aeetes for his help, and it was placed in a sacred grove with a dragon to protect it. Phryxus grew up and married the daughter of Aetes, lived a long life and was happy.

Jason grew into a strong young man, kind, passionate and very skilled. Chiron told him how his father had been long overturned by his uncle, King Pelias, and it was time for him to return to his homeland and face his destiny. There was a great feast to celebrate his leaving and promises of support and help were made between all the students. Chiron gave him a pair of solid sandals that would stand him in good stead on the long journey back home.

He walked through the mountains, the olive groves, and down to a river. There on the bank was an old woman. 'Young man,'

Aries, the zodiac sign – represents the Ram of the Golden Fleece.

she cried out, 'please help me across the river. It's too deep for me, and I am afraid I will lose my footing.'

Jason smiled, bent down, and she scrambled onto his back. The flow of the river was stronger than he had anticipated, the woman much heavier, and underfoot was more unsteady than he expected. He staggered a couple of times, and the old woman held him tighter. His foot slipped on the wet, slimy stones, his ankle turned, one of the leather straps on his sandal snapped, and floated downstream. By the time he got to the other side, the woman was so heavy to carry. She slipped off his back. Jason stood before her, wet, bronzed, and with one sandal on his foot.

'Well,' said the woman, 'truly you are the son of prophecy.' As she stood there, she changed. 'I am Hera, Queen of the Heavens. I have tested your honour, integrity and kindness.' She pointed to his bare foot. 'Be careful, young man. There are dangers ahead, but I will be with you. Have faith in yourself and the teachings of Chiron. You will be a great king.' With that she disappeared.

From the moment he left the river, over the plains and on the road to the city, whispers followed Jason as he made his way. Fingers pointed. Children dared each other to run up and touch him. He came to the city gates, and they continued to follow him as he made his way to the palace. One of the guards called out, 'The man with one sandal!' Suddenly all heads turned, leaned out of windows, all trying to see. The guards ordered him to follow them, while the crowd stood, peering into the darkness of the corridors.

King Pelias, the uncle, the usurper, was sat on his throne. As Jason was brought in, he paled.

'Who are you? What do you want?' spluttered Pelias.

Jason stood tall. 'I am Jason, son of Aeson, the rightful king. I have returned to find my father and take my place in this kingdom.'

'Your father was a fool and could not reign here – that's why I had to step in and take over. He's retired now, on some warm, cosy island. Out of harm's way.' Pelias smirked. 'We thought you long dead.'

Behind him his advisers were muttering to themselves. 'The prophecy of the man in one sandal! The oracle foretold he will bring King Pelias down!' Their voices could be heard, high-pitched and agitated. Pelias snarled at the advisors to be quiet. He turned to Jason.

'You look a reasonable young man. Perhaps you think you could be king here. I need some advice. If someone threatened to take away your most beloved possession, what would you do.'

Now, maybe Jason was naïve and did not know it was a trap, maybe he was flattered by the king asking for advice, or perhaps he was cunning in his own way – but he answered honestly and from his heart. 'I would send such a man on a quest to get the Golden Fleece, because only a true hero could do that and be worthy of such an ambition.'

Pelias laughed. 'By your own words be it. Bring me the Golden Fleece, and if you do, then I will stand aside. But I suspect that the challenge will be too much for you.'

Jason stood his ground. 'I am no fool. I have many skills, but I will need the company and support of others, and a ship to make the journey.'

Pelias laughed again. 'Take what you need, but do not bother me until you lay the Golden Fleece at my feet.' He waved his hand to dismiss Jason, but those who knew Pelias could just see his hands shake and a slight unsteadiness as he stood up.

Jason began recruiting his band of companions. Notices and heralds were sent calling for men who were brave enough to join an adventure of hardship and danger, and which would most certainly involve a fire-breathing dragon. Surprisingly there were many offers.

First was Argos. He knew the seas and how to build a ship. It would be called the *Argo*.

The mighty hero Hercules joined with his infamous strength and his giant club. Jason thought of the great deeds Hercules had already mastered. He felt overwhelmed at having this great hero amongst his companions.

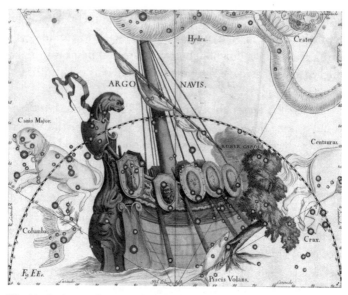

The ship the Argo can be seen in the night sky. From Greece it could be seen
straddling the both the north and south hemispheres and was one of the biggest
constellations. In the eighteenth century, the astronomer Nicholas Louis de Lacaille
broke it down into three smaller constellations – Puppis the stern, Carina the keel,
and Vela the sail.

Hera looked down from Olympus. She had her own reasons
to be dismayed at her husband's son, Hercules, joining the
voyage. Her plan for revenge would not work if he was part of the
company. Hercules had the strength and courage to undertake
the task all by himself. How would Jason manage to outshine
him, to become the king he was meant to be? She would have to
think about this.

Hercules was closely followed by twins Castor and Pollux.
They were the twin sons of Leda but by different fathers: Castor
by Leda's husband, the King of Sparta, and Pollux by Zeus.
Castor was good with horses while Pollux excelled at boxing and
wrestling. Pollux was immortal, but the love between the brothers was so great that they shared all their earthly adventures.

When Castor died, Pollux pleaded to share his immortality with his brother. Zeus denied him that, but instead set them both up in the stars as Gemini, so they did not have to be separated.

Other crew members were Tiphys, who could navigate by the stars; Lynceus, who could see great distances; Calais and Zetes, who were sons of the North Wind; and Orpheus, who could charm beast, man, sea and air with the sound of his lyre.

Orpheus is not in the night sky, but his Lyre is.

Argus built the ship for fifty men and the heroes filled the seats. There was one woman – the huntress Atalanta with her bow and arrow. When the company assembled, Argus was appointed the ship's captain, but who would be the leader of the company? Jason put it to the vote, feeling that maybe a proven hero like Hercules should lead. But the company was clear: it was Jason's mission, he should lead.

When it was time to launch the *Argo*, the ship was too heavy. Orpheus played on his lyre and enchanted the ship to move by

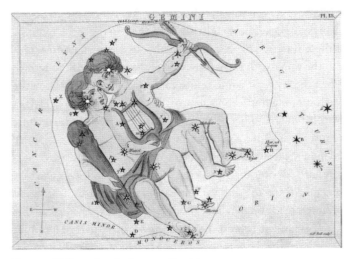

Castor and Pollux – the twins known as Gemini.

itself on land until it was in the water. King Pelias, the usurper king, watched from his palace. He shivered and wished he had done away with Jason when he first arrived.

With Tiphys navigating and the brothers Calais and Zetes creating the wind to fill the sails, they sailed off in great spirits. Jason stopped at the island where Chiron and the school were. The ship was provisioned, with many stories and words of advice shared at a great feast, and then they were on their way. After that it seemed that at each island there was danger facing them.

At one point the air was so still that not even Calais and Zetes, the wind brothers, could make a difference. With a fury, all the heroes pulled their oars. The ship moved very fast, but it took its toll until one by one they dropped from exhaustion. Hercules was the only one still rowing at fast speed until his oar caught in something and snapped. Becalmed again, they made their way to an island. While waiting for Hercules to make himself another oar, several of the heroes went collecting firewood. Among them was Hylas, the friend and companion of Hercules.

In Olympia, Hera looked down and smiled. She saw a way to both exact revenge on Hercules and to ensure that only Jason would have central role in this journey. As Hylas came across the springs of Pegae, he caught sight of the Naiads – the nymphs of those waters. They were enticed by his beauty and drew him into the water with them. Once there, he had no wish to return. When the cry went out to return to the ship, Hercules was on the beach sunning himself, his new oar by his side. 'Where is Hylas?' he called out. Another of the company pointed inland. Hercules went searching for his companion.

Up in the heavens, Hera smiled.

A haze fell over the ship, a sense of urgency to leave. The sails were raised, and the ship left those shores. Afterwards, Jason tried to remember why they had to leave so quickly that heroes like Hercules got left behind. But there was a small part of him that was relieved that he was no longer under the shadow of Hercules. He could be his own man.

There were many adventures on the way. Each time, Jason learned more about being a leader. Not everything could be achieved by just wielding club or a sword.

They stopped at Salmydessus. King Phineus greeted them and invited them to his palace. The crew were shocked at how dishevelled the king and his people were. They were offered food hidden under clothes, and invited to eat in small rooms or even closets.

King Phineus apologised. 'I offended the Gods and now we are plagued by harpies. Birds with faces of old women, who have skins so thick no sword can piece them. When we eat, they swarm and attack us, carrying off our food. There is nothing we can do but hide the food, eat little and discretely. Can you help us?'

Jason called his crew. 'Let's have a feast and see what we can do.' As soon as the spread was laid, in through the window and doors came the screeching harpies, their claws striking at the company before they seized the food. They even walked up and down the table, peering at the food to see what they fancied. Some of the company took out their swords and clubs to the birds – but to no avail. Jason nodded to the sons of the North Wind – Calais and Zetes. They summoned the force of their father, breathed in, rose in the air, and then blew. They were so powerful that the harpies were blown out of the room, off the island, across the seas, so far away that they were not able to return. Then the brothers made their way back to find another feast in place with fresh food and no harpies!

'Thank you,' said King Phineus. 'Now, I will give you some help. Near here is a channel of water that cuts through the land and will take months off your journey. But beware! It is guarded by the moving rocks, the Symplegades, that crash together whenever anything passes through, be it a bird or a ship. They move very fast, and you will not be able to sail through in time, but I will tell you the secret and give you something to help.' He whispered to Jason, who nodded. 'Thank you. That will help shorten our journey.' He took the box that the king offered him.

Lynceus, he of the far sight, was set to look for the channel of water as they passed the coastline. He gave a shout. A flock of birds could be seen making their way to a small inlet. As they got halfway through, the cliffs started to move. The birds rose into the air above to avoid them, but the sound of the clashing cliffs echoed around.

The crew were ready at their oars. Jason opened the box Phineus had given him. It was a dove. He threw it into the sky towards the channel. The dove got halfway through when the cliffs started to move. The second the cliffs crashed together, Jason cried, 'Row, row for your lives.' As the cliffs started to pull back there was just enough room for Tiphys, the navigator, to steer in the *Argo*.

'Row,' Jason said to his crew. 'Play!' Jason turned to Orpheus, who played his lyre with such fury that it charmed the sea to push the *Argo* through. The cliffs had to complete their full cycle, so the *Argo* was nearly three quarters through when the cliffs began to close again. 'Row! Play!'

But it was not enough. The rocks were moving so fast. Up in Olympus, Hera was keeping an eye on Jason. She turned to Athena, who nodded. Just as it seemed that the *Argo* might be crushed, Athena placed one hand on one of the cliffs to hold it back and with the other she gave the *Argo* a huge push. As the last part of the stern entered open water, the crew cheered. Released from Athena's grasp, the cliffs crashed together with a sound that echoed over half the world, and never opened again.

There were many other adventures before they arrived at the island of Colchis. Word of their adventures had gone before them. King Aeetes, with his daughter Medea and his court, were all waiting on the quay as they arrived.

'What do you want?' asked King Aeetes – sharp and to the point.

Jason weighed his words. 'My father is the rightful King of Ioclus. His usurper will rescind the kingdom to me if I can bring him the Golden Fleece. As the son of a sovereign king, I come to you seeking the hand of friendship to restore my family throne.'

Aeetes was concerned that Jason may try to take over his kingdom with his army of heroes. If he handed over the fleece he may be seen as weak, but he did not want to inflict a battle on his people and army. There needed to be a compromise.

'If you are worthy of this prize, then you need to demonstrate that you are as good as me. I will give you a challenge that I have undertaken myself. If you succeed, then I will let you approach the tree in which the fleece hangs, guarded by the dragon. How you do that is up to you and the Gods.'

'Very well,' said Jason, 'I agree.'

'I have a field that need ploughing, and seed set to it. Then you must reap the harvest. Once you have done all that, you may be permitted to approach the tree with the Golden Fleece. Is this acceptable?'

Jason thought this did not seem so difficult. 'Yes!' he responded.

'Good. The field is the Field of Mars, a rocky, uneven ground. It can only be cultivated by the plough made by Vulcan, pulled by two fire-breathing bulls who you must master. Then you must take these seeds and plant them. They are dragon's teeth. They will grow into fighting men of the dead. Tomorrow you will start.' King Aeetes laughed and turned away.

Jason gulped. Perhaps a little more complicated than he thought.

Up in Olympia, Hera was taking afternoon tea with Athena. They watched and they listened to Jason's latest exploit – he always provided such good entertainment. Perhaps this time he could do with a little help. Hera summoned Eros and gave him instructions. He grinned. He so enjoyed interfering with the lives of men and women.

Medea, daughter of Aeetes, waited as her father and the rest of his court made their way back to the palace. This young, handsome, bronzed Jason intrigued her with his audacity. Then – *ping* – she felt a sharp scratch. She felt an overwhelming passion for him, and a determination to help in anyway. Eros chuckled. Spot on target! Hera and Athena cut their cake and poured another cup of tea.

Medea caught Jason's attention. He knew who she was. He waited, while the words tumbled from her mouth. 'Do not underestimate the danger you are in. My father underplays the seriousness of the challenge. Let me help you – you will need protection from the breath of the bulls.'

Jason shrugged. 'I have a good team. Calais and Zetes will have my back and will use their breath to contain any fire.'

Medea shook her head. 'To be worthy of this challenge, you must do it alone. Even the slightest burn from the bulls' breath will instantly poison you. I can help you, discretely. I will make you a potion that will protect you.'

'Thank you,' responded Jason. 'But why are you helping me?'

For a moment Medea look puzzled, as if it was a question she asked of herself. Then, with clarity and self-confidence, she replied. 'You are a man of prophecy, certain to be a great king. The gods look down on us, and our fortunes are intertwined. Here I am merely a king's daughter, but my destiny is to be a queen.' They stood looking at each other – so close that they exchanged breath.

Then Medea cried out, 'Tomorrow!' turned, and left. The Argonauts put up their tents over their ship. At dawn of the next day, Medea slipped under the awning and found Jason in the bow of the ship. He stripped and she covered every single part of him with the oil she had prepared, muttering spells and incantations. Then she slipped out again, before the men awoke. Lynceus found where the Field of Mars was with his far sight. Only Jason walked to the field – the rest of the Argonauts remained on the ship. This was Jason's challenge.

The field was surrounded by Aeetes' army. A podium had been set up for Aeetes and Medea to view the proceedings. The bulls had already been released into the field, stamping their feet, the fire blazing from their noses. In one corner was the plough they were to pull. Jason took the time to study them. When Aeetes arrived, no words were said; just a herald trumpeted, and the challenge started.

Jason circled the bulls. They breathed fire. Jason walked towards the first one. It roared and everyone gasped as the flames encompassed him. It was exceedingly hot, but his skin did not burn. With all his strength he wrestled the bull to the ground, dragged it over to the plough, and hooked it up.

Jason turned back to the other bull, which was already charging. The flames covered every part of him as the bull got closer and closer. Jason put out his two hands, and although the flames danced up and down, he was unaffected. He grasped the horns of the bull, and with one almighty flip the bull was on the ground. Jason led it to the plough. The bulls were now complacent and subdued. Jason steered them and the plough to prepare the field for the seed. Job completed, Jason strode to King Aetes and put out his hand.

King Aeetes seethed. 'Someone helped you!' He threw a glance at his daughter, but she sat, steely looking ahead. He gave Jason the dragon's teeth to plant. 'Now we shall see what you are really worth.'

Jason walked up and down, giving fair distance to each seed. Then stood back. In his ear he could hear the voice of Medea: 'Stand in one corner of the field,' she said, 'and wait until they are full grown.'

Jason watched as slowly each seed unfurled from the ground, taking the form of a warrior of the dead wielding sword or battle axe. Aeetes' army was alert around the field. None of this crop would be allowed to leave. The dead army took one stride towards Jason. He was poised, sword in one hand, dagger in the other.

A voice in his ear. 'Take one of the boulders in front of you and throw it in to the middle of them.' This would mean letting go of his sword. Could he do that?

In one movement he threw his sword up into the air, crouched to grab a small boulder, gave it his best shot to hit one warrior in the middle of the dead army, then caught his sword again. In an instant the dead warrior demanded to know who had hit him, then lashed out at the nearest one. The dead army began fighting amongst themselves. Laughing, Jason pitched another boulder, and another. Heads and limbs were strewn across the Fields of Mars, sinking back into the ground. Only two were left. Jason faced them, raised his sword, and then there were none.

Jason marched towards Aeetes. 'I have fulfilled two parts of the challenge,' he said. 'Now let me approach the fire-breathing dragon that protects the Golden Fleece.'

Aeetes scowled. 'You must have had help!'

Jason shook his head. 'Can you see anyone? My crew have been at the ship. No one else here. Let me at the Golden Fleece.'

'Tomorrow!' spat Aeetes, who turned away and with his army returned to his palace.

Medea stood by Jason's side. 'My father will not let you near the Golden Fleece. He and his army will destroy you before he allows that.'

'Then,' said Jason, 'we had better be prepared. Come with me.'

They returned to the ship, and he gave orders to ready the ship to sail at a moment's notice. With Orpheus at his side and Medea leading, the three made their way through the darkness around the city, up into the mountains, though a thick wood, into a clearing. In the middle was a tree, with the Golden Fleece gleaming and lit up as if daytime. Curled around the tree was the dragon that stirred as they approached. Jason marvelled that its teeth were so powerful as to seed the dead army. He just hoped that there were no loose teeth now. As the dragon rose, Orpheus began to play. The dragon was entranced and began to sway with the music. Medea gave Jason some herbs wrapped in a small cloth.

'It's Juniper,' she whispered. 'When you can, hurl it into his mouth just as he takes a breath for the fire.'

Orpheus played. The dragon swayed, inching closer to Jason, who raised his sword, ready. In slow motion, as the dragon took breath to release the balls of fire, Jason threw the small package down its throat. In a quivering slump the dragon was soon asleep.

'We must hurry,' said Medea. 'It will not last long.'

Jason clambered over the sleeping dragon and climbed the tree. The Golden Fleece was in his hands. The three fled from the clearing. Behind them, in the distance, they heard the dragon give an almighty roar, the night sky lit up with the flames of its breath, but they were safe from him at least.

The Argonauts had raised the sails, they were ready to go. Jason turned to Medea, looked into her eyes. He said, 'Come with me. Marry me. Stand by my side as I fulfil my destiny. Be the queen you know you deserve to be.' Aeetes' army could be heard in the distance. Medea did not hesitate. They took each other's hand and leaped together into the ship.

The Argonauts sailed. The Golden Fleece hung from the mast, where it radiated light both from the sun and the moon to guide their way. Atalanta welcomed her sister adventurer.

It was a long voyage back to Ioclus. Tedious voyages along rivers like the Eridanus (now the Danube) that took them further westwards, until they finally re-entered the Mediterranean Sea, and they made the final journey home.

> The Constellation River Eridanus is in the night sky. Some people say it was the River Danube.

At the court of King Pelias, Jason presented the Golden Fleece and claimed his reward. The ageing Pelias laughed, and described how he had killed Jason's father, so that the kingdom was legally his, and destined to be passed to his own son. Jason was furious, and asked Medea to help him take revenge. Knowing that Pelias was concerned at his increasing age, she approached his three daughters and empathised with their ailing father. 'I can help,' she said, 'I can make a potion that makes him younger. Let me show you.'

She took an old ram, had it killed and placed in a cauldron of boiling oil. With words, herbs and potions, she muttered the binding words. The ram sprung out of the boiling oil as a young sheep. Excited at being able to do something for their father, the three sisters killed their unsuspecting father and placed him in the boiling oil, expecting him to spring out as a handsome young man. But no such thing happened. Jason was elated and tried to take the throne – but the people would not accept him as king with a wife who was an enchantress. And so, despite all the trials to get the Golden Fleece, the kingdom was passed to Pelias' son after all.

There were other adventures for Jason and Medea, but we leave them here as they set sail for Corinth.

Meanwhile, in Olympia, Hera was well pleased. Revenge was sweet. It had been Pelias who had defiled her temple all those years ago, and it was Medea she had wanted in the right place at the right time to exact her final revenge.

2

THE REST OF THE
WORLD: STAR STORIES

THE PROMISES

The tradition bearers I consulted advised retelling these stories from within their own culture, rather than by their relationship to a star pattern. However, I could only choose a few to be representative of the wide range of stories. There were rarely two stories that came from the same tribe or culture.

My first promise has been to group them by continent, on the basis that they share the same sky, but that does not mean they share the same culture. This is not really satisfactory, as Africa reaches from the far north of the Tropic of Cancer to the far south of the Tropic of Capricorn. It is a continent made up of many countries, each with several tribal groups with their own stories and cultures. Australia is a single country and has over 400 aboriginal groups with their separate stories and cultural mores. Where possible, I have tried to identify the community that the story has come from, but in some cases this is not recorded.

My second promise has been to retell stories that have a variety of points of view, so there are not fifteen version of hunters pursuing women in the night sky, nor ten different rabbits in the moon nor eighteen groups of hunters chasing bears in the north.

My third promise has been to provide notes on individual story sources and a bibliography of sources consulted for star stories and the context of the cultures they come from. This should help the interested reader get a head start on the research sources if they want to learn more.

The fourth promise is a table that shows the Greek names for the star patterns to which some of these stories refer.

REST OF THE WORLD STAR STORIES BY GREEK CONSTELLATION/STAR/OTHER

Cassiopeia	N. America: How Coyote Scattered the Stars (Navajo)
Corvus	N. America: How Coyote Scattered the Stars (Navajo)
Delphinus	Oceania: Irdibilyi, Wommainya and Karder (Australia)
Gemini	Europe: The Eyes of Thiazi (Norse)
Lyre	Irdibilyi, Wommainya and Karder, Australia
Orion as belt and sword	Africa: The Warrior, the Khunuseti and the Three Zebras Oceania: The Husband and Wives Who Became Stars (Australia) Oceania: The Three Brothers (Australia)
Scorpion	N. America: How Coyote Scattered the Stars (Navajo) Oceania: The Emu in the Sky (Australia) S. America: The Llama Star (Peru, Incas)
Southern Cross	S. America: The Llama Star (Peru, Incas)
Taurus the bull	Asia, Krittika: The Seven Wives of the Rishis (Hindu)
Ursa Major, aka The Plough/ Big Dipper	Asia, Krittika: The Seven Wives of the Rishis (Hindu) Europe: The Seven Stars (Armenia) N. America: The Great Bear and the Six Hunters (Seneca) N. America: Follow the Drinking Gourd – The North Star to Freedom N. America: How Coyote Scattered the Stars (Navajo)

Named Stars

Alcor (in Taurus)	Asia, Hindu Krittika: The Seven Wives of the Rishis
Aldebaran (in Taurus)	Africa: The Warrior, the Khunuseti and the Three Zebras
Altair (in Aquila)	Asia: The Cowherd and the Weaving Girl Australia: Irdibilyi, Wommainya and Karder
Antares (in Scorpio)	N. America: How Coyote Scattered the Stars (Navajo)
Betelgeuse (star in Orion)	Africa: The Warrior, the Khunuseti and the Three Zebras Australia: The Three Brothers
Polar Star, aka North Star	Europe: The Veil of Imatutan (Estonia) N.America: Follow the Drinking Gourd – The North Star to Freedom N. America: How Coyote Scattered the Stars (Navajo)
Rigel (in Orion)	Oceania: The Three Brothers (Australia)
Vega (in Lyre)	Asia: The Cowherd and the Weaving Girl Oceania: Irdibilyi, Wommainya and Karder (Australia)

Other

Milky Way	Africa: The Stars and the Star Road Asia: The Cowherd and the Weaving Girl Oceania: The Emu in the Sky (Australia) Europe: The Silver Woman Europe: The Veil of Imatutan (Estonia) N. America: How Coyote Scattered the Stars (Navajo) S. America: The Path to Abundance (Argentina, Toba Indians) S. America: The Llama Star (Peru, Incas)

Pleiades	Africa: The Warrior, the Khunuseti and the three Zebras.
	Asia: The cowherd and the weaving girl (China)
	Asia: Krittika: the seven wives of the Rishis (Hindu)
	Europe: The Seven Stars (Armenia)
	N. America: Seven Wise Men (Lenape/ Delaware)
	N. America: How Coyote scattered the stars (Navajo)
	N. America: Origin of the Pleiades (Onondaga)
	Oceania: The trials of the girls (Australia)
Northern Lights	Europe: The Veil of Imatutan (Estonia)
	N. America: The Children of the Northern Lights (Arctic and Greenland)
	Europe: Daughter of the Moon, Son of the Sun (Sami, Siberia)
Orion Nebula	Oceania: The Three Brothers (Australia)
Perseids meteor shower	Europe: The Veil of Imatutan (Estonia)

Africa

The Rabbit Prince (South Africa: Shangaan)

In the early times there was a rabbit and a duyker, a kind of antelope. They decided to grow their own mealies and calabashes. However, the duyker's patch always seemed to have been cropped by someone else. They made a trap that they hoped would catch any birds who were stealing from them. The next morning, they found a beautiful bird with long wings. They tried to capture it, but it struggled as they tried to release it from the trap and it managed to fly away. They set the trap for a second time and next day the bird was caught again. This time they took her back to their home before releasing her.

They were intrigued that she had a long feather in one wing. Rabbit guessed that the feather held her strength. He plucked it out, and the bird turned into a Sky Maiden. He hid the feather and invited the maiden to stay at his hut. She agreed, for without the feather, she could not return to the sky and her own people. Every day she was visited by birds who demanded to know when she would return to the sky. She assured them that all was well, and she would return soon. 'Have you lost your long feather?' they asked. 'The Rabbit has it for safe keeping,' she smiled.

She observed Rabbit over a few days and began to like him for his wisdom and cunning. She thought him smarter than any of the chiefs in the domain in the sky and wished he was more like her – a person of the sky. She asked Rabbit if she could see her feather. Rabbit was unsure; he was afraid she might try to escape, as he liked having the Sky Maiden with him. She took the feather, held it in her hand, and struck Rabbit. In an instant he turned into a man. Now he could take his chance to woo the Sky Maiden. It was the convention that, on betrothal, land would be given. Rabbit had no land of his own, only land he shared with the duyker. He was so desperate to win the favour of the Sky Maiden that he killed the duyker and served him on a plate in a feast. He asked her to marry him, and as she agreed, he gifted the lands to her. 'Let's keep this a secret,' she said, 'my family would not understand.'

Meanwhile, the birds were restless and blamed Rabbit for the absence of the Sky Maiden. They decided to ask the help of Mouse and Woodpecker, who were magicians. They suggested a poison. However, the Sky Maiden learned of the plot and warned Rabbit, so he was unharmed. When Mouse and Woodpecker learned who the poison was for, they were full of regret and resolved to be his friends to protect him from the birds.

As time passed, the Sky Maiden wanted to return to her family. She asked the Rabbit for the long feather, and he brought it to her. She set it on the ground. It grew taller and taller until it reached the clouds. Rabbit climbed up first, then the Sky Maiden. Mouse and Woodpecker followed – they wanted to continue to protect their friend.

As they reached above the clouds, they found they were at the entrance to a cave that was blocked by a stone. Mouse could not move it by nibbling, but Woodpecker tapped all over and found a point that made the rock swivel open. As the four stepped into the cave, a monster rose from behind a large stone. He had two great horns with a human head on each, and eyes all over his body that glared. The Sky Maiden merely took the feather, struck the monster, and he turned into smoke. The four walked into the cave and out the other side. There was a valley as green as the one they had left on earth. They watched as birds flew down from the sky and changed into human form. 'My family,' whispered the Sky Maiden, 'the people of the clouds.'

'Who is this man?' demanded her father.

The Sky Maiden responded, 'He is a man of the earth. He is a wise and cunning as any chief here. I choose him as my husband.' Her father and his people were furious. This could not be allowed. The Rabbit must be killed! They pretended to welcome the Rabbit and his friends and had a feast to celebrate, but instead they planned to poison him. While the food was being prepared, Mouse kept an eye on everything. He saw two dishes put aside, and a powder sprinkled on them. He warned Rabbit not to eat anything.

The people of the clouds were not pleased their plan had failed. The next day, the Sky Maiden's father asked Rabbit to visit another Kraal across a plain in the sky. Aiming to please, Rabbit did as he was asked, but when he was halfway there the skies opened and hailstones pelted down. With no shelter, this looked like the end. 'Do not fear!' came a voice, and it was Woodpecker at his side. 'Get down and I will cover you with my wings. My magic will keep you safe.' She spread her wings as the hailstones fell about them.

The sky people were upset that their plan had not worked. Instead, they arranged a hunt for all the men to attend which would last for several days. Plenty of opportunities for an 'accident' to happen. But Woodpecker heard everything. Using her magical skills, she made a portion of fat of mamba, fat of python and the skin around the lungs of a tiger. These would give protection. She placed them in three bags and gave them to the Rabbit with a warning to be careful.

Whilst the hunt was successful, the attempts on Rabbit's life were thwarted at every turn. When they returned from the hunt, he went to the Sky Maiden and told her everything that had happened.

'They will not rest until I am dead. I must return to my own world. It is up to you whether you stay here or return with me.'

The Sky Maiden knew her own mind. 'I will return with you to the earth. Summon Mouse and Woodpecker – we must leave now.'

She drew out the long feather and pointed it towards the earth. Again, it stretched until it rested next to Rabbit's house. One by one they climbed down the feather, sliding the last few feet.

The Rabbit decided that the Sky Maiden had given up so much for him that he wanted to create his own tribe and community for her – she could not just live in a hut. The Woodpecker smiled. 'In the three little bags are the potions that will enable you to gather an army and to amass many cattle.' And how he did that is a story for another day.

The Warrior, the Khunuseti and the Three Zebras
(Bushmen: Namaqua)

Pleiades, Aldebaran, Orion's sword, Orion's belt, Betelgeuse.

Tsui, the Sky God had seven daughters called the Khunuseti. They were all married to the man they called the Warrior. One day, the Khunuseti wanted the Warrior to bring them three zebras. As a proud warrior with seven wives, he was pleased to do this. He got close to the zebras and shot his arrow. But he was overconfident, and his arrow missed the three zebras. He tried to retrieve the arrow without disturbing them, but he became aware that there was also a lion watching the same prey. Now he was trapped. Unable to get his arrow, unable to move because of the lion. He was ashamed to go home to his wives without any meat. So he shivered in the darkness, all alone.

The Khunuseti or the 'Stars of the Spring' are the Pleiades. The warrior is the bright star Aldebaran (in the constellation of Taurus), his arrow is the sword of Orion, and the three zebras are the belt of Orion. The lion is Betelgeuse.

The Stars and the Star Road (South Africa)

Milky Way.

In the early days of the world, when the sun went down the sky was dark with no stars nor lights. When man learned to make fire, there was a spark of light and warm in front of him, then it died down and was gone.

One night, a girl was sitting by the fire, warming herself. She took some of the ashes and blew them into the air, watching the flames of the fire reflected in the falling ash. She took some green bushes and put the wood on the fire. It burned differently. Bright sparks went here and there. She stirred with a stick then took up some of the ash from the ground, stirred the greenwood again, and when the bright sparks appeared she threw the ash across the fire. Ash and sparks melded and flew up into the sky. They made a bright road across the sky. We call this the Milky Way.

Then she took some roots she had been eating and threw them into the sky. The old roots become stars that gave off a red light, whilst the young roots just gave a golden glow. As they shone, they sang for the girl.

One star grew and grew and the sang the names of all the other stars. He became known as the Great Star. Every night all the stars walk across the sky, either side of the Star Road, until it is time for them to go just as the dawn wakes and makes a pathway for the sun. When he comes, he brings the brightness for all the people to enjoy their work and play.

ASIA

The Cowherd and the Weaving Girl (China)

Vega, Altair, Hegu 1 and Hegu 2
are in Aquila, Pleiades, Milky Way.

This is known as the Chinese valentine story. It dates from the Zhou dynasty (1046–256 BCE), so it is at least 2,000 years old.
   There are many variations with the key characters and actions being the same.

Zhinu was the daughter of the Jade Emperor of the heavens and his wife – the Queen of the Heavens. Her father doted on her and made plans for her to marry. But she did not want to submit to her father's will – she had already fallen in love with Niulang. He was also a god, but his duties were to care for the celestial cowherds. In the ranking of the gods, he was very low. They kept their love a secret until the day that the Jade Emperor announced in his court that Zhinu would marry one of the earthly mortal kings to seal his allegiance.

Zhinu refused. The Jade Emperor was incensed that his daughter would deny his will in public and demanded to know why she refused to comply, threatening that he would strip her of her immortality. The whole court held their breath, fearful at this change of tone of the emperor. Zhinu said nothing – afraid of that harm would come to her lover if she spoke his name.

Then Niulang stepped forward and declared his love for her. The Jade Emperor took one look at him, and roared so loud that they even heard him on the earth. With one sweep of his arm, Niulang was gone. The court gasped. Zhinu fell to the floor, weeping and wailing. The mortal king, whom she was to marry, took one look at her and turned away. He knew she could never love him. The pain and sorrow of that day would stay in her heart, and she could never share it with him. Marriage was about allegiance, but not on these terms.

The days were long for Zhinu. She joined her six sisters in the sky, weaving the clouds into many patterns. Sometimes she wept for Niulang, and the tears fell as rain onto the world below.

But the Jade Emperor did not have the power to kill an immortal – only to strip immortality away. Niulang was reborn to a mortal family; poor, but loving and kind. He grew up with no memory of what had happened before, but only with a sense of longing for love. While he was a small boy his father found an old ox that seemed to have no home. He gave him to Niulang to look after. Everyone was amazed when he seemed to instinctively understand about the needs of the old animal. The two became inseparable, and the family joked that they didn't know who was looking after who, even when Niulang grew to be a young man.

Meanwhile, in the heavens, Zhinu never forgot Niulang. As her sisters shared their beds with their husbands, she heard their tales about the sweet delights of marriage. But it was not for her.

The seven goddesses remained in the sky, weaving the clouds, until one of them suggested that they visit the Earth and take pleasure in bathing in one of the lakes. The Jade Emperor's palace did not have such natural pleasures. Secretly they came down to the earth and found a place of such beauty of water, rock, tree and flower that they could see it was worthy of a goddess's attention. They removed their dresses and sank into the warm water, which flowed over them. They swam, they floated, and they splashed. Such fun they had and for the first time in a long time Zhinu laughed.

Niulang was in the cow burgh with the ox. He heard laughter, and it stirred a memory inside him. He stood up and tried to work out where it came from. Then his old friend the ox bellowed. And spoke for the first time.

'Niulang,' he said. 'You have looked after me all these years, and in turn I have looked after you. I have kept you safe until this day. Come with me.' The ox had been one of the gods in the Jade Emperor's palace when Niulang was banished. He had felt that the Jade Emperor was wrong in what he had done, but knew he dare not say anything. So, he had decided to come to the earth as an ox and search for Niulang.

The ox told Niulang to ride on its back. Niulang refused at first, thinking the ox was too old, but finally consented. With a god's speed, the ox covered the ground to the place where the laughter could still be heard. As they arrived, Niulang caught sight of the maidens singing and swimming. His eye was drawn to one of them. Then he gasped. The longing for love, that he had felt all his life, overwhelmed him and he fell to his knees. He could feel the blood rushing through his veins. No words came from his mouth. There was a warmth in his heart and an ease he had never felt before.

Zhinu had seen the ox and its rider and called out to her sisters that a stranger was amongst them. Her sisters fled for the shore to collect their clothes and make their way back to the heavens, calling to her to join them. But Zhinu shook her head. There was something about the young man that felt familiar, so she walked towards him.

As she got closer, she could see the young man on his knees, shaking. But she could see beyond his body, into his true soul, and in that minute she knew she had found Niulang. When she was within touching distance, all the old memories came flooding back to him. He stood up, arms outstretched, and embraced her.

So many words. So many kisses. So many memories. So many promises.

Her sisters implored her to rise back to the heavens with them, but she said no. She wanted to remain on the earth with her mortal Niulang. The sisters shook their heads. They would not tell the Jade Emperor where she was, but if he asked, they would have no option but to give an answer. Zhinu responded that she had spent so much time pining for her love, she did not want to spend one second more without him. The sisters rose back into the sky to weave the clouds.

We know them as the Pleiades.

Niulang and Zhinu were wed and lived with their ox. Niulang was very successful and became a wealthy man raising oxen. The old friend ox always had pride of place and was pleased to allow their two children to ride on his back, and swing on his tail.

But the Queen of Heaven noticed that her younger daughter was missing and demanded to know where she was. The sisters could only put her off for so long before they revealed where she could be found.

The Queen of Heaven was furious, came down to earth and in an instant, stole Zhinu away, and put her among the stars for safe keeping.

We know the star as Vega.

Distraught, Niulang cradled his two children and asked the ox what he should do.

'I am old in this earthly body,' said the ox. 'Slaughter me with all the right honour and prayers, then take my skin off. Wrap it around yourself and the children and I will use what powers I have left to rise you up into the sky.'

Niulang and the children wept at the thought of losing their friend but wept more at the thought of losing their lover and mother. With all the right honours and rituals, the ox was slain, and the skin taken off. Wrapped around himself and the two children, Niulang did not know what to expect – but instantly they began to rise into the air to become stars themselves, close to their mother. They were so pleased to be reunited.

Niulang is the star we know as Altair and the children are the stars Hegu 1 and Hegu 2.

But the Queen of Heaven was not going to be outwitted by a mere mortal. There was no way that she would allow the two to be together in the night sky. She created a river of stars that pushed the two stars for the lovers far apart. They could no longer see each other.

This river we call the Milky Way.

But true love will always find a way. This story of the unrelenting love came to be known on the earth, and on the seventh night of the seventh moon, all the magpies rise up into the sky, and they create a bridge between the two lovers so that, for one day a year, they can meet.

Altair and Vega (in Lyre) are part of the asterism we know as the Summer Triangle. The two children are part of Aquilae known as Hegu 1 and Hegu 2.

Krittika: The Seven Wives of the Rishis (Hindu)

Plough/Big Dipper, Pleiades, Alcor, Taurus.

The gods Brahma and Savitiri had seven daughters, known as the Krittikas, who each were happily married to the seven Rishis – sages or wise men. Each day they helped to bring up the sun.

The Rishis can be seen in the sky as the seven stars we know as the Plough (or Big Dipper).

One day, in their wisdom, the Rishis offered a sacrifice to the god of fire, Agni. When the rites and rituals were over, the Rishis dispersed to attend to their tasks. But Agni was now in this world and wanted to explore. He saw the wives of the Rishis. He became infatuated with them and wanted to possess them. They were faithful to their husbands, and Agni knew that his thoughts were improper. He decided to transform into a fire in their household to observe them. He did this for a long time. Eventually, he was overwhelmed with being close to them and not able to express himself. He turned away into the forest to work out what he could do.

As he did that, he was seen by Swaha. She had always had a secret passion for Agni, but he had made it clear that he was not interested in her. Now she saw a way to get what she wanted. Using her powers and enchantments, she changed herself into the image of one of the wives. She followed Agni into the wood. He was very agreeable to spending time with her as one of the wives. She was delighted, and they tumbled together. As he rested, she changed herself into a bird so that no one could see her and flew away. At the top of a mountain, she found a lake, then took the semen from inside her, and placed it there. She then took the image of another wife, then another. Agni was overwhelmed – he hadn't realised that the wives of the Rishis would be so welcoming to him. He tumbled with each one as they appeared to him.

Every time, Swaha placed the resulting semen in the lake. However, Swaha could not assume the shape of the seventh wife. She was too pure for the magic to take shape. From the seed from six encounters with Agni, a child was conceived with great powers: Skanda, who grew into a man within five days.

Rumours flew about that the six wives of the Rishis were the mother of Skanda. Things had been seen and not understood. The six wives were driven out and divorced from their husbands. Eventually, Swaha went to the king and confessed that she was the mother of Skanda. The king told the Rishis that their wives were innocent, but they continued with the abandonment of them.

The six sisters rose back into the sky as the Pleiades. But the seventh sister – she kept away from the scandal and rose in the sky as the star we know as Alcor, shining bright in her purity.
Swaha is one of the stars that we know as the horns of Taurus.

EUROPE

The Eyes of Thiazi (Norse)

Gemini.

Idun, the Goddess of spring and immortality, came to live in
Asgard with her husband Bragi, God of poetry and song. With
her she brought a basket of apples that she gave freely to the gods
of Asgard, which they found brought youth and energy to them.
They would age in years, but their bodies and minds were as agile
as the day they ate their first apple. The marvellous thing was that
no matter how many apples she gave away, there was always a
fresh number the next day. She was content to remain in Asgard
while her husband attended to his minstrel duties.

Now Odin, King of the Norse Gods, with his friend Honir and
Loki, took delight in exploring Midgard, this mortal earth, with
the intention of finding new wonders. One day they found themselves very hungry. Loki, always trying to please Odin and get
into his favour, promised to find food. He soon came across an
ox. He looked around, and as he could not see anybody nearby,
decided that it belonged to no one and that it was his to take.

He took the ox back to Odin and Honir, slew it, and then built
a fire to roast it. The three rested, joked amongst themselves, then
hearing their bellies rumble, looked forward to the meat. After
two hours, Loki put out the fire and pulled off the meat. This was
long enough for it to be well roasted. But it was raw. Embarrassed
that it was not done, Loki reset the fire and let the beast roast for
a further hour. But it was still raw. Desperate for an explanation,
he looked around. In a tree nearby there was the biggest eagle that
you have seen. It laughed.

'I have enchanted it so the fire has no effect,' it said. 'But if you
agree for me to take first share, then I will release it and freshly
roasted meat will be on your tongue.' The three men were so

hungry they would have agreed to anything. So, Loki reset the fire again, and this time the meat was well cooked. Loki pulled it off the fire and stood back to allow the eagle to take the first share. The eagle rose above the tree, stretched its wings, and then swooped down, taking all the meat with it.

Loki was furious. He grabbed a nearby stake and thrust it at the eagle as it reached up to the sky. But to his astonishment, although the stake pierced the side of the eagle, he found he couldn't let go. Some enchantment kept him close. The eagle let go of the ox and rose into the sky. The meat fell at the feet of Odin and Honir, who greedily seized it. But Loki now found himself battered against the side of the mountain, dragged over the surface of a glacier, and scraped across the briars and brambles on the ground – unable to let go and fearful that his arms were being drawn from their sockets.

'Mercy,' he cried. 'I'll do anything for you. Just stop.' Three times he cried this, and then the eagle finally rested.

'I will only stop if you promise to bring the Goddess Idun and her golden apples out of Asgard.'

With that, the eagle turned back into his true shape. A giant named Thiassi. 'If you don't,' he said, 'I will pluck you out of Asgard myself.'

'I agree,' said Loki, as he tumbled back to the earth. 'I promise.'

Odin and Honir didn't notice Loki when he staggered back to the fire. They had no interest in what he had to say and assumed that Loki had chased the eagle off while they ate the roast ox.

Loki said nothing, but when they returned to Asgard he started hanging around Idun. When Bragi was on one of his minstrel trips, Loki saw his chance.

'I hear,' Loki said, 'that on Midgard there are some very fine apples that look better than yours and taste so much sweeter.' At first Idun ignored him, confident that her apples had no match. But his wheedling words began to work their way into her confidence, so that in the end she demanded that he show her these new apples.

'Oh,' he said, 'you'll have to go to Midgard, and bring your basket so that you can compare.' Irritated at the audacity of Loki to suggest anything less about her apples, she agreed. Together they walked out of Asgard down to Midgard.

'Well,' said Idun. 'Where are they?'

In the distance, Loki could see an eagle flying towards them. 'Oh, we will find them quite soon, just over there.' Loki pointed to a group of trees. Idun peered at them and did not see the eagle until it had grabbed her in its claws and taken her up into the air. Ignoring the screaming that still rang in his ears, Loki turned and walked back to the bridge at Asgard.

Thiazzi, the eagle – now as the giant – placed Idun in the finest room of his mountain fortress. He asked her to stay with him, to become his wife, and to let him eat of the apples of immortality. On each demand she turned him down, so her room became a prison, her only solace a narrow-slit window that looked down over the mountain to the lakes below and up to the sky. The whole world before her but denied access to all.

Meanwhile, in Asgard, as time passed Idun's absence was noticed. Perhaps, they thought, she had accompanied her husband for the first time on one of his minstrel tours? The Gods were beginning to feel the ravages of age – not just the drooping skin, and the weakening of the hand, but the tricks of the mind where the memory of the morning had been lost after the midday sun. No need to worry – when she returned, one apple each would restore any youth lost. It wasn't until Bragi returned that they realised she had disappeared. Odin, with his all-seeing eye, struggled to find her in Asgard or Midgard.

Where was Idun? Odin sent his ravens to seek her. They found her on the other side of the narrow slit in her tower. There was nothing to do but to report back to Odin. Thor, when he heard, wanted to attack straight away. He would use his hammer to smash the tower down. But Odin wanted to be more strategic. He didn't want outright war with the giants. Instead, he summoned Loki.

There were rumours that Loki was the last to be seen with Idun, but it was so long ago no one was sure anymore. Odin only knew that it would take someone sly and a trickster to bring back Idun. Loki fitted the profile perfectly.

'I will,' said Loki, 'but I need Freya's falcon cloak first.' Freya, Thor's sister, reluctantly handed over her cloak, making all kinds of threats as to what would happen if the cloak wasn't returned pristine. Loki thanked her, then issued instructions that wood and resin wood shavings should be placed by the wall that led to the entrance to Midgard.

With the cloak on his back, Loki strode across the floors of Asgard and was transformed into a falcon that soared high into the sky. Without a look behind him, he flew straight to the fortress, and following the instructions of the ravens, found the slit window, and got himself through. Idun saw the bird on the floor in front of her and, worried that it might be hurt, stooped down to pick it up. But Loki fluttered away from her, muttered some runes under his breath, turned both Idun and her basket of apples into a nut, then grasped them in his claw to make good his escape again.

Thiazzi knew as soon as Idun left the fortress. He raced up to the top part of the fortress, and with his eagle eye spotted the falcon and its precious cargo. He turned himself into the eagle and flew to regain his prize.

In Asgard, the gods were watching the skies and cheered when they saw the falcon. The cheers faded away when they saw the eagle behind, swiftly catching the falcon up. Would Loki manage to get back before the eagle got him? Freya muttered under her breath about what she would do to Loki if he harmed her falcon cloak. Whether her words did any good, no one knows, but from somewhere Loki found an extra spurt of speed and pulled away from the eagle. As the falcon flew over the walls of Asgard, Freya called, 'Now.' Every god then placed a torch to the wood piled in readiness and a wall of fire roared up just as the eagle approached. Unable to stop, Thiazzi was overwhelmed and burned in the flame and fell in the grounds of Asgard, where the Gods slaughtered him.

Idun was transformed but traumatised, not really clear how she had ended up in the grasp of Thiazzi. But her apples were fine, and she shared them with the Gods, who felt their new-found strength coursing through their limbs, revitalising them.

Odin stood by the corpse of Thiazzi. He knew there would be recriminations with the giants. Some diplomacy and compensation would be needed. And a gesture towards the greatness of Thiazzi for his other deeds. Odin bent down and took the two dead eyes of Thiazzi, held them as shining orbs in his hand and then threw them into the night sky.

We see them now and they are known to us as the twin stars of Gemini.

The Silver Woman (Siberia: Sami)

Milky Way.

In one of the Sami villages in Siberia lived a young woman who was so fast on her feet the wind just about caught up with her. She knew her own mind, and when it was time to be married she was determined it would be of her own choosing. Only a man who could be as fast as her would do as a husband.

When potential suitors came, she would run far away, and none could catch her. However, there was one young man who was determined to try and trained hard. When he approached her, she ran away. He ran after her. She ran through the forest. He kept after her. She even turned around to check if he was still there. He was. She smiled. She ran over the plains. He was still there. She ran over the hills. He was still there. She ran up a mountain, and he followed. At the top she took a great leap and vanished into a low-hanging cloud.

The young man reached the top of the mountain exhausted

and collapsed to the ground. He struggled to breathe. The young woman looked down from the cloud and stepped back onto the mountain.

'Are you alright?' she asked. 'What can I do for you?'

He looked up at her, and gasped, 'Water?'

She looked around. There were no springs nearby. What to do? She brought her breast close to his mouth and squeezed it until milk ran from the breast to his lips. But the wind blew hard, and the milk flew from her breast up into the sky and into the heavens above, where it created the Milky Way.

Aghast, the young woman let go, and some of the milk ran over her. In a moment she herself was turned into a block of silver. The young man was horrified that the object of his affection was now a silver statue. No matter how much he pleaded with the wind, nature and the gods, the young woman remained a silver statue. He kissed her lips and then left to go down the mountain, over the hills, across the plains, back to the village.

There he told them what had happened. They said prayers for the silver maid, but the young man could not be reconciled, and he died of a broken heart.

The Veil of Imatutan (Estonia)

Milky Way, Northern Lights, Sun, Moon, North Star, Perseid meteor showers.

Imatutan was one of the seven daughters of the heavens. She was responsible for the care and well-being of the birds who migrate – making sure there was enough food and shelter for them as they arrived in the north on their long journey from the south. She loved this role and did it with care and diligence, even using the winds to blow dust into the eyes of any hunters. Imatutan was very beautiful and many of the stars wanted to marry her – but she was a woman who knew her own mind.

First came the North Star in his dusky coach with six black horses and ten gifts. He was very shy and uncertain whether a beautiful star like Imatutan would consider him for a husband. 'Would you marry me?' he asked.

Imatutan smiled. 'I am sure that I can trust you as a husband,' she said, 'but you will always be there in the sky, never moving. Everything circles around you, and you will be looking over me all the time. I am a free spirit, as your wife I could not stay in the same place all the time.'

The North Star was sad and went back to his place in the sky, where he stays in his watchtower from which he has not moved to this day.

The next suitor arrived in a silver carriage drawn by ten grey horses. Imatutan was excited to see who had come to woo her. It was the moon, and as he stepped out of the carriage his white light fell around her. With him he had brought twenty gifts made of silver.

'My dear,' she said, 'I am flattered by your attention, but I suspect I may never know who you are. At one point you are a silver crescent, then a full-bodied roundel. Who would I look to for love? Most of all, I realise that you need another for your light and existence. It doesn't come from within you, and I know I cannot give you that light myself. I must decline for your sake.'

The moon was sad but understood her words and left, taking all his gifts with him.

Quite soon after a carriage of gold came, driven by twenty gold red-maned horses that shone brightly. She was not surprised when out of the carriage came the sun to ask for her hand and showered her with thirty gleaming gifts.

Again, Imatutan smiled. 'You are most certainly the hottest in the sky, but I watch you each morning. You rise in the east and travel the sky to the west. Day after day. Sometimes you go a little higher, and sometimes you go a little lower, but it is the same path each day. I cannot be constrained to always take the same road every day. I love the changing seasons. My spirit is too free for that.'

The sun was sad, sank a little lower in the sky, but departed, taking with him all his gifts.

Time passed. Would there ever be a suitor who would give her what she wanted?

A diamond carriage arrived, drawn by a thousand white horses. The carriage stopped. For a moment nothing stirred and Imatutan held her breath. Could this be the one? Then, out of the carriage, stepped the Lord of the Northern Lights. His green and blue hues scattered here and there, drew together in a spike and then rushed away again in a dance exploring all the corners. 'Oh yes,' said Imatutan, 'you are the right partner for me. You do what you will, where you will, there is no holding you down. I can see that you will be the right soulmate for me.'

The Lord of the Northern lights concurred with her and gave her gifts of diamonds and jewels. They agreed the marriage would happen as the birds flew south. Together they celebrated their newfound love. As the sun rose to begin his journey across the day sky, the Lord of the Northern Lights arose and told to Imatutan to prepare for their marriage on his return.

Thrilled, she made all the preparations for a wedding day. She dressed herself in a white flowing dress, and on her head she wore a long veil studded with the diamonds and jewels the Lord of the Northern Lights had given her. And she waited. Spring passed. Summer came and left with the birds. The snow settled on the earth. No word from the Lord of the Northern Lights. He did not come. Had she misunderstood him? Would he come another day? But as the sun made many more crossings, followed by the moon, under the North Star she knew that the Lord of the Northern Lights would not be returning.

She grew sad and remorseful. She could not think of anything but the Lord of the Northern Lights. She yearned for what might have been. She even neglected the birds under her care, and as they returned from the south there was no food or shelter. The birds were devastated. Some of them appealed to her father in the heavens that Imatutan had forgotten them. He was horrified at what had happened, and summoned the spring winds to lift her up to the edge of the sky above with her long veil stretched out behind her. He used the diamonds to fix her into the firmament. With the last stud in place, Imatutan opened her eyes. With the change she felt different – not so sad now. She could watch over all the earth and guide the birds to their winter quarters, where there would be shelter and food. She stands there still as the Milky Way.

From there she cares still for the birds, and they follow her line in the sky as they fly from the south to the north and back again. In August of each year something amazing occurs. When the sun is lower in the sky, the Lord of the Northern Lights returns with his 1,000 horses. Imatutan's heart pumps with joy – finally she can see him. Now she understands the cycle of the earth, but she cannot be with him because she is now fixed in the sky. He is so changeable that he cannot keep a promise but for a few short weeks they can celebrate their love. That is the time she takes some of the diamonds from her veil and scatters them over the earth to show her love has returned.

We call them the Perseid shower of meteors.

The Seven Stars (Armenia)

Pleiades or plough.

Once there were six brothers, all very handsome. The youngest went hunting. He was out all day and felt tired, so he rested by the side of a brook.

A young girl came there to get some water. She saw the young man and gasped – this awoke him. In that moment when their eyes met for their first time, the magic of true love began to weave a path between them, and they expressed their undying passion for each other.

'Let's get married!' said the young man.

'Yes!' said the young woman. Then, mindful of her duties to her parents, she added, 'But I must take this water home first.'

The young woman carried the water, careful not to spill a drop – she was so excited. When she arrived home, she told her mother and father that she was to be married. They were shocked. This is not what they wanted for their daughter. Using an enchantment, they took her away to a far and distant place. At the same time as they compelled her, the young man also fell asleep.

It was several years before the family felt it safe for her to return, feeling confident that all thoughts of the young man were gone from her head. However, the first time the young woman walked down to the brook to get some water, she saw the young man still there, sleeping. She called out his name and he awoke. The enchantment had been powerful, and he turned into an angel and rose into the air. Suddenly he stopped, fell to the ground, and died. The young woman was so distraught that she died by his side. Her family blamed themselves and allowed the two lovers to be buried together.

The five remaining brothers, who had searched for their young brother high and low, were so distraught at his loss they killed themselves to be with him. The six brothers and the young woman were raised to the heavens where they remain to this day.

Some say they are the stars known as the Plough, some as the Pleiades.

North America

The Great Bear and the Six Hunters (Seneca)

Plough/Big Dipper.

> There are many indigenous stories that link Bears to the
> Plough/Big Dipper.

Six men went hunting, but try as they might, there was no game. One man said he felt sick, but really, he was just tired and wanted to take it easy. The other men were concerned and made a litter of poles and a blanket. It took four men to carry him as well as their own packs. The last man had to carry everything else, including a kettle. They were hungry and walked for days. Eventually they found bear tracks, but they looked several days old. Their bellies rumbled – it was better than nothing. They tracked for several days until the paw marks looked fresh. The bear was close.

The men dropped the litter with the 'tired' man on it and raced to catch up the bear. The 'tired' man didn't want to get left behind, so he jumped up from the litter and chased after the men. Because he had been resting, he had more energy than the others. He soon overtook them, caught up with the bear, and killed it. The other men were still making their way towards them when they began to lift from the ground. They were so eager to get to the bear that they did not realise until they reached the sky and could look down and see the 'tired' man and the dead bear. Even then they did not stop, and eventually rose into the night sky. You can see them there today.

> The men make up the stars of the Plough/Big Dipper. The man who
> carried the kettle is in the bend of the Plough/Big Dipper, the kettle
> is a star in the handle and a small one nearby. The bear is at the
> outside corner. At first frost, drops of oil can be seen on leaves of the
> oak tree, and this is the oil and blood of the bear.

Follow the Drinking Gourd: The North Star to Freedom

Plough/Big Dipper, North Star.

Strictly speaking this is a song and not a folk tale. But it is included here as an example of how the stories of the night sky made a difference to people's lives. At the turn of the eighteenth and nineteenth centuries, when slavery was prevalent in the south of the US, the only way to find freedom was to go north. Whether they were guided by a 'conductor' or had to make their own way, without maps or compasses, finding the Pole Star for north was crucial. The asterism in the Great Bear of the Plough (UK) or the Big Dipper (US) was the key. They called it the drinking gourd, as that was the shape they were familiar with. Following the line from the side of the drinking gourd gave a sight line to the Northern Star. The 'map songs' gave the directions in a form that was easily remembered. The second line is a covert message to leave at the winter solstice. It would take a year to get to the north part of the country, by which time the Ohio River was frozen and they could walk across. At other times of the year, it would be impassable.

Follow the Drinking Gourd
When the sun comes back and the first quail calls,
Follow the Drinking Gourd.
For the old man is awaiting for to carry you to freedom,
If you follow the Drinking Gourd.

The river bank makes a mighty fine road,
Dead trees to show you the way,
And it's left foot, peg foot, traveling on
Follow the Drinking Gourd.

The river ends between two hills,
Follow the Drinking Gourd.
There's another river on the other side,
Follow the drinking gourd.

Where the great big river meets the little river,
Follow the Drinking Gourd.
For the old man is awaiting for to carry you to freedom
If you follow the Drinking Gourd.

I thought I heard the angels say
Follow the Drinking Gourd.
The stars in the heavens
Gonna show you the way
Follow the Drinking Gourd.

Seven Wise Men (Lenape/Delaware)

Pleiades.

> The people of the Lenape are believed to be the grandfathers of all the Algonquin people. 'They do not own the land, they are of the land, they belong to it.'

A long time ago there were seven men of the Lenape'wak. Over the years they had learned many things, and between them they were very wise. People would come from far and wide to ask their advice or to learn from them. The responsibility became too burdensome for them, they felt they had no time to themselves.

They held a great council and decided to turn themselves into rocks. That way no one would find them. They would have time to do what they like best – to commune with each other and learn even more. They walked far beyond the villages and turned themselves into rocks.

All would have been well, but one young man found the rocks and became fascinated by them. He would spend day after day looking at them. One day he spoke to one of the rocks. The young man was mighty surprised when the rock spoke back to him.

'You are the seven wise men!' he cried.

'Yes,' they said, 'but please don't tell anyone. We wish to be left in peace.'

The young man agreed on the proviso that he could still come to see the rocks and ask questions. 'Alright,' they said, 'but don't tell anyone.'

Well, he was true to his word – he told no one. But in the village, it was noticed he kept disappearing, and one day he was followed. Thus it was that the seven wise men were rediscovered. Yet again, they were deluged by people asking for advice. 'This is enough,' they said. 'We must get away from here.'

In the twilight they changed their form back to men, then walked through the night until they came to a beautiful valley, where they decided to rest and become cedar trees. No one would find them now.

However, being glorious trees in a beautiful valley meant that there were visitors to enjoy the sights. The people realised again that these were the seven wise men. And again, they were deluged with people asking for advice and help. The seven wise men were desperate to have their privacy. Where could they go? They had been men, rocks and trees. Where else could they go?

The creator – Kishelamakank – took pity upon them. He scooped them up and placed them in the night sky, where they could be seen by all but not disturbed or annoyed. These stars are called Asiskwataja'sak. They twinkle in the night sky with all their knowledge and are only disturbed by the prayers that waft up to them.

These are the stars we know as the Pleiades.

The Children of the Northern Lights
(Arctic and Greenland)

Northern Lights.

A hunter worked hard to build up his winter stash of meat that he kept in a cave. One day he returned from hunting to find that some of the meat had disappeared. He was furious! This was his lifeline for the winter. He decided to hide and punish whoever was stealing his meat.

He hid in the back of the cave and waited. After two days he was about to give up, when he heard a noise. A group of children suddenly appeared in the cave and began pulling on the meat to take it away. The hunter stood up, his spear ready to wreak his vengeance. The children turned to look at him. He realised they were the Northern Lights children – the spirits of stillborn children who dance with their afterbirths in the sky. He could not remember seeing the Northern Lights for some time.

One of the children stepped forward.

'May we have some of your meat. We have been so hungry that we are unable to dance in the sky. Unless we can have some food, we may never dance again.'

The hunter was full of compassion for the children. 'Please take what you need.'

That night the Northern Lights returned to the night sky.

How Coyote Scattered the Stars (Navajo)

Pleiades, Polaris, Plough/Big Dipper, Cassiopeia, Corvus, Scorpion, Milky Way, Antares (star in Scorpion).

The Navajo people have a long tradition of telling tales of the stars. Resting in their *hogans* or homes, they tell the stories of the creation of the night sky.

The hogans are built according to the sky plans given to them by the gods when they emerged from the underworld. They start with a frame of five poles: one to the north, one to the west, one to the south, and two to the east with a door to welcome in the dawning sun. The poles are placed in the ground as the sun makes its progress – first to the east, then the south, the west, and finally the north pole, which holds them together. This is the proper way to enter a hogan: sunwise. Then they cover the poles with tree limbs and earth.

In the beginning time the first hogan was made in the same way, except it was covered with sunbeams and rainbows. It was in this hogan that the Navajo gods of creation met to plan how to make the world and what to put in it. They created the sun and moon and put them in the sky – placed them so that there was day and night. But even with the moon in the night sky, it was too dark.

As they contemplated what to do, Black God, the god of fire, joined them.

On his ankle he carried a small group of stars called Dilyehe, forged from fire. The other gods asked him what they were. Black God did not speak but instead moved sunwise around the hogan. At the south pole he stamped his foot. Dilyehe jumped to his knee. At the west post, Black God stamped again – this time Dilyehe jumped to his hip. At the north, stamp! And to his shoulder. Then finally to the east – stamp! And Dilyehe jumped to his left temple. And there it stays until today: in ceremonies, the representative of Black God always wears the stars on his left temple.

The other gods were greatly cheered. These bright stars would light up the night sky. They asked Black God if he could make more to make the night sky beautiful. Black God brought out a pouch made of fine faun skin and from it he drew out a single bright crystal. He held it so that everyone could see, then he placed it in the night sky – exactly in the north. They called this the North Fire, which never moves and guides travellers. All the other stars would turn around this star.

He opened his bag again, and this time took out seven pieces of crystal. He placed them near the North Fire.

'This,' he said, 'will be Revolving Man, as it moves around the North Fire.'

Then he took out more stars and placed them near the North Fire. 'This will be Revolving Woman. Both will forever circle the North Fire.'

The creation gods nodded in agreement.

Black God then took out five shining crystals. He turned to the east and placed them in the sky – they called it Man with Feet Spread Apart.

Continuing sunwise to the south, he placed more stars into the sky and created First Big One, and below he placed three smaller stars he called Rabbit Tracks, as that is what they looked like in the snow. Turning westwards, he placed Horned Rattler, the Bear and Thunder. With great joy he placed little groups of stars here and there. The night sky was shining with the lights. Black God made a copy of the stars on his forehead – Dilyehe – and placed them too. He took a step back to look at his work. The creation gods gave a sigh of satisfaction. Finally, Black God reached into his bag and brought out thousands of little stars and carefully placed them across the sky: Which Awaits the Dawn. It was beautiful. The creation gods nodded their heads and agreed this was excellent work.

Then Coyote appeared. A trickster, who asked awkward questions and brought chaos to the carefully laid plans of the gods. 'Why have you started without me?' he demanded.

The other gods said nothing. Black God looked Coyote in the eyes. 'I have placed stars in the sky to guide and protect men and women. The patterns give rules to live by, and if they follow them, they will be peaceful and content.' Black God sat down cross-legged, his faun-skin bag under his foot. Coyote could see that he was protecting it. He bent down and snatched it away before Black God could stop him.

Looking into the bag, Coyote could see more sparkling lights. 'Lovely!' he said. He reached in and took out a large one, then took the bag and threw it up into the sky. All the other crystals flew into the sky, going here, there and everywhere.

'I had them in patterns, in order – everyone could see them!' cried Black God.

Coyote laughed. 'And now they have chaos! They can choose for themselves which path they follow!'

The gods started to murmur and mutter. Once the stars were in the sky, there was nothing more they could do.

Then Coyote coughed. 'Of course,' he said, 'there is one thing missing in the night sky.'

The gods looked amongst themselves. What could he mean?

'Me – I'm not there!' The gods were aghast at the thought of Coyote in the sky. But Coyote opened his hand to reveal the last large crystal. Then, very carefully, almost like Black God had done, he placed it in the sky in the south.

'That will be my own star!'

He stood back to admire his handiwork. 'Now it truly is beautiful!'

The Navajo say that as humans it is our task to find the harmony between the ordered and predictable world of Black God and the unpredictable world of Coyote. And as Coyote says, we can choose which path to take!

The Dilyehe is the Pleiades; North Fire is Polaris; Revolving Man is Big Dipper; Revolving Woman is Cassiopeia; Man with Feet Spread Apart is Corvus; First Big One is the upper part of Scorpius; Which Awaits the Dawn is the Milky Way. The Coyote star is known in our system as Antares – and is the bright object in our constellation of Scorpius, and one of the largest stars visible to the naked eye.

Origin of the Pleiades (Onondaga)

Pleiades.

In the beginning of time, there were hunters travelling across the land seeking good hunting ground.

They reached Kan-ya-ti-yo – a beautiful lake with grey rocks surrounded by great forest trees. They could see the land was plentiful; there were many fish in the lake. They watched the deer come down in flocks to drink at the lake, and they could even see bears on the far shore. This was a good place to settle.

The people came together and gave thanks to the great spirit of the water for providing them with a bountiful hunt and all the gifts they needed to be happy. They began to build their lodges there, settling in for the winter months. They sought the game they needed to fill their stocks for the winter months, when there would be little available to them.

The children played their games, but most of all they loved to dance. One boy would mark the beat on a hollowed-out tree trunk, and the rest would dance, trying to outdo the others with the most outrageous moves. Then one day an old man approached them. He was dressed in white feathers and his hair was as white as the clouds. He was a stranger and, as he came near, the drumming stopped and so did the dancing.

'That's right,' he shouted. 'Stop dancing. This is no place to be dancing. The spirits are not at rest. If you continue to disturb them, then bad things will happen.'

The children looked at him, wondering what to do. Then the boy with the tree drum started to beat it, and with a laugh the children danced again.

The old man came several times again – always saying the same thing. But the children just laughed at the stranger. He was not of their tribe and did not know their ways. After all, they had been dancing for some while and nothing had happened.

As the days drew into winter, the children still danced, but now they were hungry after they finished. They decided to ask their parents for some food so that they could assuage their pangs of hunger with a feast. But food was tight and strictly rationed, and their parents told them not to waste food at a feast, but to stay at home and just eat there instead.

The children were disappointed, but decided that dancing had become more important for them, they would continue regardless.

They danced and danced, some feeling dizzy and lightheaded from their pangs of hunger.

Suddenly, one child felt so lightheaded that his feet left the ground and started rising to the sky. He called out to the others. One by one they rose. Some were scared. Some thought it was a big adventure. 'Don't look back,' called out one of the boys.

The children rose higher and higher.

An old woman had seen the children in the air and ran to warn their parents. Fearful of what they might find, they gathered up some food to bring to the children, hoping that would tether them to the earth. What they saw horrified them. They called up, trying to entice the children back with promises of food.

One boy was tempted and turned to look down. As soon as he did that, he became a falling star and fell to the earth. The other children kept looking skyward, out into the heavens, and there they became stars twinkling and forever dancing.

The people of the Onondaga call them Oot kwa tah.

We know them as the Pleiades.

OCEANIA

The Husband and Wives Who Became Stars (Australia)

Orion's belt.

The widow sat beside the body of her partner in life and cried out to the stars and the ruler of the heavens. 'Why did you take my husband? He lies in the dust. I am left here alone and without children to comfort me. No one to help me and care for me in my old age.' Such was her outrage and despair she took a sharp-edged flint and began cutting herself, watching the blood drip to the ground.

Nepelle, ruler of the heavens, looked down and his heart was moved at the woman's anguish. He called to Nurunderi, his servant. 'Here is a lost babe of the heavens. Take him and leave him where the widow may find him. He can give her the solace she needs.'

Nurunderi took the babe and placed it under a bush close to the widow, and stood back and waited. Soon the babe began to cry. The widow in her sorrow did not hear him at first, then the cries cut through to her heart and she swiftly found him. She gave her thanks to the heavens and called the child Wyungara – meaning he who returns to the stars.

As the child grew, he brought much comfort to the widow, and in turn she gave him everything that a mother can. Supported by his foster uncle, the widow's brother, he learned all the ways of the people and fulfilled all the rituals and tasks for the initiation ceremony. His foster mother was proud. Now a man, he could support her in her old age.

Wyungara was a fine hunter and one time, when out in the bush, he heard an emu's call. He searched and instead found a young woman. He told her his name, and not to be afraid. She smiled. Her name was Mar-rallang and she had heard of his prowess. The two of them talked with great ease until the sun started to set, and then they said their farewells.

The next day, Wyungara went hunting again in the same place, hoping to meet the young woman again. This time he heard a swan's cry, and while searching for it he found another young woman. He introduced himself and she smiled and said her name was Mar-rallang too. She was sister to the woman he had met the previous day, and they were so alike that were called 'two in one'. They also talked with great ease until the sun started to go down and they left for their homes.

Wyungara then spent many days with one or other of the sisters. He knew he wanted one of them to be his wife – but which one? He decided to ask them if they would both be partners in life with him. To his joy, they both agreed. However, his uncle – the widow's brother – said that this arrangement was not right according to tribal law, and he must have only one wife.

Wyungara and his wives decided that they wanted to live together, and so Wyungara bid farewell to his foster mother. In the darkness the three of them left and travelled a great distance to where there was no one else living, and they set up camp there. The three of them were very happy.

Then one day Nepelle, ruler of the heavens, saw them. He realised what had come to pass and was outraged. He called Nurunderi. 'Wyungara is a son of the heavens. Why is he married to not one but two daughters of earth? This is forbidden!'

Nurunderi squirmed. 'He was only a baby when he was left on the earth. No one told him he was a spirit of heaven.'

Nepelle was furious. 'This cannot be allowed. Separate them!'

In his spirit form, Nurunderi came to the earth and found Wyungara's camp. He set fire to the bush surrounding the camp, hoping that would drive them apart. Wyungara awoke, the smoke choking him. He grabbed his longest spear, and then, taking one wife under each arm, he sprinted and jumped over the ring of fire. But the fire had spread, so he jumped again and landed in the lake. The reeds that were in the lake were dry and caught fire too, and soon the roaring flames were covering the whole lake.

Wyungara and the two Mar-rallangs looked at each other in love and in fear. Then Wyungara took his long spear, braced it against the bed of the lake, and then told his wives to climb over him, to the top of his spear. Then, commanding all his might, he threw the spear into the sky and watched as it sped across the stars. Exhausted, he fell back into the water and waited for the fire to consume him, knowing he had done all he could for the ones he loved.

Nepelle was shocked. This wasn't what he intended. Touched by Wyungara's last action to save his wives before himself, he relented. When the spirit of Wyungara left his body, Nepelle raised it up into the heavens to the place where the spear had taken the two wives – and all three became stars shining in the night sky, united for ever.

You can see them in the sky – we know it as Orion's Belt.

The Three Brothers (Australia: Yolngu)

Orion, Orion's Belt, Orion Nebula, Betelgeuse and Rigel.

Many of the fish and animals in indigenous Australian culture have a prohibition where there is a season to eat and a season not to eat. Sometimes it's to protect the new young, and sometimes it's to protect all the fish. It encourages the people not to become over-dependent on one food source.

Once there were three brothers from the Yolngu tribe, who went fishing in their canoe. It seemed the fish did not want to be caught that day, and the only fish they caught was the kingfish (which people today call the yellowtail amberjack). However, the boys were from the kingfish clan, and it was forbidden to eat that fish.

At first, they threw the kingfish back into the sea. But their bellies were crying out for food. The youngest boy said, 'I must eat something. We have caught nothing else. This must be a sign that we can eat this fish.' The elder one said, 'You cannot eat that – it is against all of our tribal law, it is the fish of our people.' Neither of them noticed the third boy, who had seized a fish and was starting to eat it raw. The other two brothers watched him, astonished. When nothing happened to him, they fell upon the remaining fish.

When they finished, they did not speak a word. They were ashamed that they had broken one of the traditions of their people. As one they plunged their hands into the water and sprinkled their face to remove the debris. They hoped no one would know what they had done. They began to make up stories of the fish they were allowed to eat and convinced themselves that this indeed was what happened.

High in the sky, Walu, the sun woman, saw all. She understood the traditions. She knew why they must not eat the kingfish. The boys must be made an example, so that everyone understood the importance of the tribal law.

As the boys made their way back to shore, suddenly they were caught up in a giant waterspout, that raised their canoe high out of the water. The boys held on tightly, uncertain of their fate.

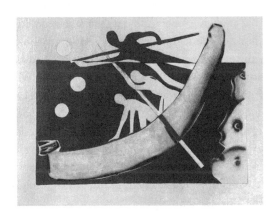

They looked up in the sky and saw Walu, sun woman, with anger on her face, her arms outreached and words pouring from her mouth. The waterspout went higher and higher, the canoe held the boys right at the top and soon they were amongst the stars. They looked down at the disappearing earth, felt their full bellies, and regretted that they had broken the traditions.

> *The constellation we know as Orion is the canoe, with Betelgeuse and Rigel as the bow and stern of the canoe. Orion's belt is the three brothers and Orion's nebula are the forbidden fish.*

The Emu in the Sky (Australia: Papunya, Northern Territory)

Milky Way.

In the Dreamtime there was a man who had lost his sight in a hunt. He convinced himself that the only thing that could cure his blindness was eating a raw emu egg. He called his wife and told her to bring him some. She looked far and wide and did her best to find them. But he got angry with her because he felt that the eggs she found were too small. He said that she wasn't trying hard enough to help him.

She tried hard to find larger eggs. Then she saw the biggest tracks of an emu that she had ever seen. The eggs of that bird must be surely big enough for her husband. She followed the tracks to the nest, and sure enough there was a very large emu, and she could see the eggs. They were the finest eggs she had ever seen. She did not want to harm the emu, so she tried throwing some rocks at it to distract it away from the eggs. But never underestimate a mother guarding her offspring. The emu stood up and charged at the wife. She tried to turn away, but the emu was too big and too determined. The weight of the emu was enough to push her to the ground. The emu used its great toe claws to lash out at the woman, who screamed. Nothing in the night heard her. She was killed instantly.

In the meantime, the blind man was feeling hungry. His wife had not returned, and he felt sorry that he had been so angry with her. As the cold of the night fell upon him and she had still not returned, he decided to go and look for her. He felt his way around the camp and found a bush with some berries on it. To quieten the moans of his stomach, he ate some. To his amazement, he could now see. Remembering his old skills, he put together some spears and a woomera and made his way into the bush.

He followed the tracks his wife had left and came across her body. He could see the blood and the broken bones. He knew it was an emu that was responsible for his wife's death and grew very angry. He heard a noise behind him. He spun around. There was the emu.

He took aim with his spear, and let it loose. The emu fell to the ground. Dead.

In the Dreamtime, many things can happen. The man who was blind no more bent over, picked up the emu, and threw it into the sky, where you can see it today as one of the dark spaces of the Milky Way.

This is a different kind of star story in that it is the gap between the stars that is identified as a creature. It can be seen in the Milky Way from the southern hemisphere. The llama story from Peru uses the same 'gap'.

Irdibilyi, Wommainya and Karder (Australia: Torres Strait)

Altair, Vega, Lyre and Delphinus (the dolphin).

In the Yamminga times, a woman named Irdibilyi lived with her husband, Wommainya, and her brother, Karder. Because he was a slow-moving man, Karder was known as the lizard. He liked nothing better to stretch out in the warm sun and sleep. He didn't hunt for himself at all. Wommainya did all the hunting for the family, including Karder, because this was the way of the tribal law.

Wommainya's favourite food was the *bardi* – grubs found at the root of the tree. While Wommainya was out hunting, Irdibilyi would take their two sons, and with her digging stick she would search for bardi. Her two sons would try to copy her, but just found honey and ants, which they ate themselves. Wommainya was always pleased to get the bardi.

There was a season when the bardi were hard to find, and Irdibilyi had to walk long distances to find them. Her two sons were with her and each time she found some grubs, the boys would try to eat them. 'These are for your father,' she would say, 'go find your honey and ants.'

All the way back to the home camp the boys were calling for the bardi. She prepared the meal and yet again the boys called for the bardi. 'No,' she said, 'these are for your father. If you want some, go and find your own bardi.' With that she turned her attention to the fire.

When the food was nearly ready, she called the boys – but they did not answer. She was surprised, and asked Karder to go and look for them. He was reluctant to get up from his cosy place but did as she asked. He looked here and there but didn't find them. He thought to himself that the boys would soon return anyway, so why should he look any further, so he returned to the home camp.

Now Irdibilyi was worried. It was tribal rule that the boys should never go outside the home camp by themselves, they were so young. What would she tell her husband?

Wommainya returned with a fine kill – a wallaby. He was pleased with himself, but he became very angry when Irdibilyi told him what had happened. He left the camp and began looking for tracks. He could see Karder's tracks, and that he had not gone very far. He felt angry. Then he found the tracks of his sons. He followed them for a long way until he came to the Ngaiyanup Water. There, in the middle of the lake, were his two sons, with water up to their necks. When the boys saw their father, they called for help. Wommainya took his beard and stretched it out across the water to reach the boys, and they grabbed it thankfully and began to pull themselves in. But the water was so deep and began to suck the boys into the lake. The boys were not strong enough to resist and they fell back into the water and drowned.

Grief and anger overcame Wommainya. He took his beard into his mouth and spat it out, again and again. He ran back to the home camp.

Irdibilyi saw him on the horizon and knew that her sons had died. She was overwhelmed by grief and welcomed the death that tribal rule demanded. When Wommainya stood before her, she rose, and he plunged his spear into her heart. She joined her sons in death.

Wommainya turned to Karder. 'You did not search for my sons, you are lazy, you don't hunt and now you are nothing to me. Leave my hearth. The rule of our tribe does not allow me to kill you. But you are not worthy to sit with men, sit only with the women.'

With his sons and wife lost to him, Wommainya died with grief.

And in the night sky you can see their story. In the stars known as Vega you can see Wommainya stretching out his beard to his sons – the two bright stars south of Vega in a constellation known as the Lyre. Irdibilyi is the star known as Altair and Karder, banished to sit by his sister away from the company of men, is in Delphinus (the dolphin).

The Trials of the Girls (Australia)

Pleiades.

In the early days there were seven young girls of an age to be initiated into the ways of the tribe. They agreed that if they were to be mothers of warriors then they must know the trials a warrior has to endure in his initiation, to prepare their sons to become the best warriors they could be. They wanted to know how to use their minds to deal with physical pain and hunger, and how to control their fears. This, they believed, was the best way to support their tribe. They presented themselves to the elders of the tribe and made this request.

The elders considered this. 'We could set you tests that would enable you to do this. What you ask is only for men.'

The girls refused this. 'Very well,' said the elders, 'we will set the tests as we would a man.'

They were told to live apart from the rest of the tribe for three years. At sunset and sunrise, they would be given a small portion of food – sometimes fish, emu or wombat. Each girl felt the hunger, the pain, and dealt with her fears of living in such a way.

After the three years were up, they felt ready to return to the tribe after their trial.

'No,' said the elders, 'this was the training. Now we have the testing. You will fast for three days while we travel to the testing place.'

Some of the girls were surprised by this and thought they had done enough, but the others encouraged them to continue. It was a long and difficult trek through the rough and thorny bush under the blazing sun. They were hungry and fatigued, but with their trust in each other, the girls persevered.

On the evening of the third day, they came to their camping ground. Food was prepared for them, and they slept. In the morning they were given a flint knife and a freshly roasted kangaroo. They were told to only cut meat just enough for their needs. No more, no less. The smell was so good that some were tempted to take a more generous portion, but then they saw how moderate the portions were that the others took and did the same. As they finished their meal, the elders congratulated them on their restraint.

'This is the first of the tests. There are other appetites you need to control – not just hunger. Each test will be harder than the last.'

They were then taken to a sacred place and told to lie on the ground. For each girl, an elder knelt before her and told her to open her mouth. With a stone axe and a pointed stick, the elder knocked out one of each girl's teeth, exposing the nerve.

'Do you feel pain?' asked the elders.

'Yes,' said the girls.

'Are you willing to have another tooth knocked out?'

'Yes,' said the girls.

The elders did the same thing again, knocking out a second tooth.

'Do you feel pain?' asked the elders.

'Yes,' said the girls.

And a third time. Pain? Yes!

At the end of the day, the girls were asked – do you want more testing? As one they said, 'Yes.'

The next day they were taken to another camping ground. There the girls were made to stand in line. One of the elders of the tribe came up before them, and each in turn he slashed the girls across the breast. The blood flowed and it was painful. Then another elder came and rubbed wood ash into the wound, this increased the pain but also allowed the wound to heal. At the end of two days the wound was starting to heal.

The elders asked, 'Do you want more testing?' As one the girls said, 'Yes.'

They were taken to another camping ground and told to rest. There was no fire. No moon. As they slept on the ground, they could feel things creeping across them. This was a test of fear. One sat up, trying to brush things off her, but the others murmured words to calm her and each other. They were sleeping on a bed of ants. All through the night there was the quiet rustling as the girls shifted in the darkness, but they did not scream, they did not shout. They held the fear inside themselves and helped each other to be brave. When the sun rose, they had remained silent and motionless through the night.

There were more tests. On the completion of each one, the elders asked, 'Do you want more tests?' As one the girls said, 'Yes. We will use our minds to conquer pain.'

Their noses were pierced, and a stick put through to keep the wound open until it healed.

They lay on a bed of hot cinders and endured the pain.

Each time they would be asked if they wanted further testing, and each time the girls would respond with, 'Yes. We will use our minds to conquer pain.'

They were taken to another camping ground. While the campfires burned, they were told the stories of the Bunyip and the Muldarpe – tormenting spirits that came in many forms – and the ghosts that frequented the area. Some of the girls shivered and trembled as they listened, but none turned away. Then the elders told them that their camp for the night was a burial ground. The girls settled to sleep, some still thinking of the stories they heard. All through the night the elders crept around the camp making weird noises and strange calls. In the morning they were calm and happy, knowing that they had controlled their fears.

This was the last of the tests, and now they could be initiated into the tribe. For such an event the elders sent invitations to the other tribes to a *corroboree* – a meeting – to celebrate.

The girls stood before the corroboree and said, 'We have endured the test of appetite, pain and fear to know what the sons of this tribe need to face. As girls we become women, and call upon other girls not to be afraid to undertake this initiation yourselves. You will learn that happiness comes of thinking of others, not yourself. Greed, pain and fear are the result of thinking too much of yourself and not others – so you must learn to control these. We call upon you girls, who will soon become women, will you do this?'

The girls and women of the corroboree cried, 'Yes!'

The Great Spirit was very pleased with what the young women had achieved and how they had inspired other girls. He decided that they should be spared the death and suffering of this life and chose to raise them straightway up into the sky to be a symbol of their sex and give inspiration to others.

You can see them in the sky – and we know them as the Pleiades.

South America

The Path to Abundance (Argentina: Toba Indians)

Milky Way.

In the early days the earth had moved, so there was a deep gorge that separated two parts of the living bush. A tree was caught by its roots on one side. It stood tall at first, but the force of the movement was so strong that it tumbled and rested over the gap – very precariously balanced. Nature had made a way for man, animals and other creatures to use that tree for free passage between the two parts of the bush. Over time, it began to creak and move. It was not as safe as it used to be.

One morning a cenin bird, a sandpiper, saw that the tree was beginning to slip. Crossing it would be dangerous for all creatures. He tried to weave some ropes to steady the bridge.

A group of young girls came by, looking for something to eat. They had been up since sunrise and could find nothing, so they decided to rest under a tree. One of the girls continued walking and found the edge of the gorge. She looked down into it – it made her feel very giddy. Looking up to the other side, she

gasped. On the other side she could see that the trees and bushes were bountiful with fruit. That was the place to be.

She ran back to her friends and told them. They were excited and the thought of food made their bellies rumble. They looked for a way to cross the bridge and found the tree that had fallen across. Although the tree seemed to be moving, there were palm leaves which the girls thought they could hold on to as they crossed. The cenin bird called out to them to wait. He would secure the tree for them. But the girls were impatient and would not wait. They began to climb onto the tree, grasping the palm leaves to steady themselves. Laughing and shouting, they began their way over the tree, anticipating what food would be on the other side. However, heavy with the weight of all the girls on the tree, it began to creak and then break. The girls screamed and shouted, trying to hold on to the tree as best they could but to no avail. All the girls fell into the gorge and died. No one could cross that way again.

The jabiru bird (a stork) arrived. He was going to the other side of the gorge. He was surprised to see that the tree had gone.

'Why did you let it break?' he asked the cenin.

The cenin shrugged. 'I was trying to mend it but it broke by itself. I don't know what to do. You seem to be a skilful person – could you fashion another bridge? '

The jabiru preened himself. 'Yes, I can do that,' he said proudly.

He used his beak to fell the trees and cut them up into long boards. He placed them across the gorge and used a palm tree to secure them and provide a handhold for any who wanted to cross.

He called the cenin to show him his handiwork. Together they walked across the bridge. It was safe and secure.

'Now,' said the jabiru, 'let us make a path so that all will know it is here so that they can pass to the place where there is an abundance of food.' He set fire to the bush to clear it, and when the smoke had settled the sparkling lights of the ash were in the sky, so that anyone can follow it and never feel hungry again.

The path is the Milky Way.

The Llama Star (Peru: Incas)

Southern Cross, Scorpion, Milky Way.

In the beginning there was peace in the land. Mankind and animals lived contentedly side by side. Then mankind became greedy, and to feed their greed they became very cruel. They would fight, steal, neglect their crops – and worst of all, they would not worship their gods. In all the world there was just one place where mankind was good and kind to themselves and the animals. This was in the High Andes.

There were two brothers who lived there. They were farmers, and the pride of their herd were two llamas, a mother and baby. Over time the brothers realised that the llamas were not eating their food. Something was disturbing them. The brothers changed the food – but it didn't seem to matter. All the llamas wanted to do was gaze at the stars.

Finally, the brothers asked the llamas, 'What is happening? What is wrong?'

The mother llama looked at them and answered. 'Now that you speak to me as one of your own, I can tell you. The star gods are warning us. Mankind has been too greedy and cruel. The gods will send a punishment, a flood, to wipe out all the badness. You must take your family and your animals higher up the mountain. For your kindness you will be spared.'

The two brothers did just that. They found caves higher on the Andes and prepared them for the day when the floods would come. Somewhere comfortable and safe for themselves, their families and all the herd.

When the rain came, the brothers led their families and herds to the caves, stocked with food, water, blankets and plenty of firewood. They lost count of the number of days the rain fell. Sometimes the brothers ventured down the mountain to see how high the water was. They could see other farms and buildings that were swept away. They tried not to see the bodies of men and women drowned as they tried to flee. They wept for the animals that died, trapped in the corrals that mankind had built. When it reached their original home, and it disappeared under the waters, they returned to the cave, and they all prayed to the gods to be spared. They prayed hardest to Inti, the sun god.

The water grew higher and started to lap at the cave entrance. The brothers did their best for their families and animals, trying to get them onto higher ground. Their stocks and provisions were getting low, they were very afraid. And they prayed. Somehow it seemed as if the mountain always grew a bit more, to be just high enough above the water.

As the water ran around their feet, the brothers had almost given up and were resigned to their fate – when the rain stopped. The water about them began to ebb, and they watched as Inti, the sun god, came out and began to warm the earth. The water evaporated and ran away. The mountain shrank down to its original size.

The two brothers and their families returned to what was left of their homes. They rebuilt them and filled it with their children and grandchildren.

The llama mother and baby, as a reward for their care of the brothers, were taken up into the sky, where even today they are revered as gods. At night time, when mankind is sleeping, they come down to the earth and drink the water out of the ocean to stop it rising again.

Llamas have a long memory. Whilst mankind has spread out across the world, they remember the long days and nights of the flood, and always keep to the high ground, just in case.

The mother llama is represented by Scorpius, the baby by the Southern Cross – and both are in the Milky Way.

THE REST OF THE WORLD: MOON AND SUN STORIES

AFRICA

A Home for the Sun and Moon (Nigeria)

Way back in the beginning of the world, Sun and Moon and Water all lived on the earth. They each had a *kraal*, a stockade, suitable for their needs, which allowed them to do their various tasks in the world. Now, Sun and Moon were good friends and loved to visit Water. They would talk about their day, this exciting new world, and let off steam for a bit. Water would talk about his family, and Sun and Moon were a little envious as they had no other family.

One day Moon said, 'We visit Water all the time. But he never comes to our homes. Is there something he doesn't like about us?' Sun said he wasn't sure. But Moon was so worried about it that Sun began to see it Moon's way.

'I agree it's a bit rude,' said Sun, 'for Water not to visit. Next time we are there we should insist that Water visits.'

And that is what they did. Water laughed. 'I would love to do that, but where I go, so do all my family! I fear your kraal will not be big enough, even if you combined both your kraals.'

Sun was a little annoyed that his friend did not say yes straight away. In fact, he felt a little insulted. 'Then we shall break down the stockade between us – there will be plenty of room! Come tomorrow!' Water was not sure about this, but Sun insisted. So, Water agreed.

Sun and Moon watched as Water flowed towards their house. They opened part of the stockade and Water trickled past them, covering all the ground. Within Water you could see fish and sea mammals. He had brought all his family. Then Water started gushing in, and spread over into Moon's kraal.

Soon, Water was up to the knees of Sun and Moon.

'Are you sure that there is enough room for me and *all* my family?' asked Water.

Sun looked at Moon, and Moon looked at Sun.

Sun was too embarrassed to admit that maybe this wasn't a good idea.

'Plenty of room,' he said, as both Sun and Moon climbed onto the roof of his hut to get out of Water's way. Waves started to form, with the white crests as it moved around both kraals.

Then Water called all the streams, rivers, lakes and oceans. The house filled up, and Water even started to cover the roof.

There was only one place left to go. Sun looked up at the sky, and as Water lapped close to the top of the hut, he took a great leap into the sky.

'Come on Moon,' he called out.

Moon was very afraid because deep down he knew this was all his fault and he shouldn't have been so suspicious. So, he waited until Sun had travelled past the mountains, and as the Water began to cover him, Moon took a leap into the sky. Thus, the sun and moon remain in the sky – where the moon follows the sun.

Asia

The Buddha and the Hare (Sri Lanka)

In the life before Buddha was born to be Buddha, he was known as Bodhisattva. He was a hare in a wood that was sumptuous and fruitful for the creatures that lived there. He was known to be very wise, and his best friends were a monkey, otter and jackal. They looked to the hare for their moral and spiritual guidance.

At that time, people who sought a spiritual path and withdrew from the world were called ascetics or *brahman*. They begged food from other people so that they could spend all their time carrying out their holy responsibilities. It was believed that giving alms and food to them was a scared duty. Bodhisattva told his friends that when a brahman asked for food, they should share the food that they had.

It was a good day for the friends. The otter came across some fish on the bank of a river. They had been landed by a fisherman who left them while he fished for more. The otter called out, 'Do these belong to anyone?' but the fisherman was too far away and did not hear. The otter was pleased with his catch. The jackal found a hut, and inside there was a lizard and a jar of milk curd. He called out, 'Do these belong to anyone?' When no one answered, he took them to his den. The monkey found some mangoes in a tree and collected some for himself. But Bodhisattva, the hare, only found grass and knew this was not enough for any brahman. 'I would have to sacrifice myself,' he thought, 'to enable a holy man to feed.'

Now Sakra, Lord of the Devas, King of the Gods, saw all this and determined to test each animal. He disguised himself as a brahman and came to the otter for some food. The otter offered him the seven fish. Sakra was pleased and told the otter he would come again the next day. He then visited the jackal, who offered him the lizard and the milk curd. Again, as the brahman, Sakra said he would come the next day for it. And similarly with the monkey.

Sakra was pleased that the animals followed the teachings of gifts to brahman, but now he would test the hare.

He came to Bodhisattva as the hare and asked for some food. The hare told the brahman to light a fire. Sakra did as the hare asked him. Then, when it was hot and burning bright, the hare turned to Sakra and said, 'I have nothing for you to eat but myself!' and with that the hare jumped into the fire. Sakra smiled and was pleased. With a wave of his hand the fire was cold, and the little hare was surprised to be alive. Sakra revealed himself to be Lord of the Devas, and praised the hare for the sacrifice he intended to make. In recognition of this, and as an example to others, Sakra painted the likeness of the hare on the moon, to forever remind everyone to honour the holy ones.

Chang E: Goddess of the Moon (China)

The Jade Emperor is the ruler of heaven, one of the highest-ranking gods and the very first of the Chinese emperors. His palace is in the heavens, where he lives with other immortal beings and those who serve them. He is greatly moved by things of beauty, believing his own greatness provided the inspiration, and the skills in making it. There was one porcelain vase in particular that he admired, and he would request that it be brought to him, so he could reflect on their shared glory. One day a girl called Chang E brought it to him, but in her nervousness, she dropped it. Even in the heavens a dropped vase still breaks. As the vase shattered into pieces, the Jade Emperor roared, and Chang E was immortal no more. She was transformed into a babe born to poor peasants on the earth, with only the faintest memories of her previous life.

Chang E grew into a young woman with pale silky skin, hair black as night and lips like cherry blossom. She was courted by many young men, but her heart went to Hou Yi, a young archer, and they were married. Inspired by her, Hou Yi trained to be the best archer in all the world. He had a bow made of tiger bone, and his arrows were tipped with dragons' claws.

At this time the world was still young, and the Jade Emperor had ten grandchildren that were suns that took turns to take the journey across the sky, illuminating the earth and providing warmth and succour.

One day the suns could not decide whose turn it was to rise. So, all ten suns rose together. The heat scorched the earth. The light was relentless; the night sky had vanished and all life on the earth was threatened. The water evaporated, crops shrivelled, the plants and animals began to die. Monsters arose from the shadows into the light and began to slaughter humanity, who could no longer see them.

In the Jade Emperor's palace there was great concern, but no matter what the emperor said, the ten suns still shone.

On the earth, Hou Yi called to the Emperor that if he could not control the suns, then he, Hou Yi, would shoot them down to save the earth. The Jade Emperor, with a sad heart, agreed.

With his bow of tiger bones and arrows of dragon claws, Hou Yi hunted the monsters down until none were left. Then he climbed to the highest mountain and called to the ten suns to leave the sky, but they took no notice of him.

Faced with no other choice, Hou Yi raised his bow to the suns. By his side was Chang E. She shielded his eyes from the blinding rays of the suns while he used his other senses to shoot his arrows at the suns. Nine arrows. Nine suns fell out of the sky. The tenth sun was so scared of Hou Yi's strength that it fell to earth of its own volition and hid in a cave, refusing to come out. This plunged the earth into cold and darkness. All that lived asked the remaining sun to come out again – but he was too scared to hear anything. Then the rooster added his own voice, calling, 'Brother, rise, rise up!' The shrill voice cut through the cacophony of sound, gave comfort to the sun, and he emerged from the cave.

The wife of the Jade Emperor, the queen of the heavens, was pleased that all was restored. She gave Hou Yi an elixir of immortality as a reward so that he could rise to the heavens.

Hou Yi was touched by this. He wanted to share it with his wife, Chang E, but first he wanted to say farewell to their friends. He left the elixir in the house they shared.

However, an earthly man called Feng Meng heard the stories about the elixir. He was also an archer and aspired to be as great as Hou Yi. He decided that he would seize the elixir of immortality for himself, so he would be even greater that Hou Yi. He broke into the house and searched for it. He was disturbed by Chang E. As soon as she saw him, she knew what he wanted. He took hold of her and forced her to reveal where the elixir was hidden. She resisted, but eventually had to concede. It was in a place where only her small hand could reach. With Feng Meng holding her arm across her back, and in great pain, she reached in to get it out. She could not bear to think of the damage that Feng Meng might cause if he was an immortal, so as she withdrew her hand with the vial in it, she flipped the lid and drank it all herself.

Now each morning it is the cock's crow that welcomes the sun rising.

In an instant Chang E ascended to the heavens as an immortal. Just then, Hou Yi arrived home, saw her rising, and realised she had taken the elixir. He was so angry that she had not waited for him. He felt betrayed and tried to shoot her down.

Watching all this was a rabbit who, like Chang E, had once been an immortal and was now a rabbit in this world. He had seen all that had occurred, and as Hou Yi shot his arrows, the rabbit hurled himself against Hou Yi again and again to disrupt his aim, until Chang E was far above in the sky.

Feng Meng was long gone by the time Hou Yi entered the house. But there was enough debris for Hou Yi to work out that there had been an intruder and a struggle. He was devastated that he had not trusted Chang E and had tried to shoot her.

Distraught to be on the moon and immortal without Hou Yi, Chang E called upon the rabbit to help her make another elixir of immortality for her husband. The rabbit rose to the moon, and she searched for herbs while the rabbit pounded them to powder with his back legs.

Not knowing the truth of what happened, Hou Yi invited Feng Meng to work and train with him. Feng Meng was so full of jealousy and bile that he tried to kill Hou Yi several times. Naively, Hou Yi took these attempts as 'over-enthusiastic' training. Bitter and twisted, Feng Meng cut a stick from a peach tree and bludgeoned Hou Yi to death.

The Jade Emperor in his palace saw all that had come to pass. He was horrified that his gift had led to Hou Yi's death. As the last spark of life was about to leave Hou Yi's lips, the Jade Emperor made him immortal and raised Hou Yi up to be one with the sun.

Chang E is the moon, Hou Yi is the sun, and they are the yin and yang of the night sky.

On the mid-autumn festival on the fifteenth day of the eighth lunar month (roughly in September or October) Chinese people still put out food to honour Chang E in the moon.

In 2020 the Chinese people named their rocket to the moon Chang E in honour of the moon goddess.

Chu Cuoi: The Man in the Moon (Vietnam)

There was once a young man called Chu Cuoi who chopped wood in the forest. He carried his axe with him, but as he walked along the path, he found a tiger cub in his way. Chu Cuoi could see it was dying and as much as he wanted to comfort it, he knew that the mother tiger must be in the area. Afraid for his safety, he climbed a nearby tree. In the distance the mother tiger could be heard calling out. She soon found the tiger cub, licked it, and pushed it as if to wake it up. There was no movement. Chu Cuoi knew that the cub must be dead, and he braced himself for the tiger's roar of grief. He was concerned that he might be caught in her anger if she could smell him.

He was surprised to see that that she did none of that. Instead, she padded over to the banyan tree next to where Chu Cuoi was hiding. The mother tiger reached up her paw and pulled down some leaves, which she put in her mouth and chewed. Then she returned to the body of the tiger cub, nuzzled it, then placed the chewed leaves in its mouth. Within a few seconds the tiger cub began to move, get up onto all four paws, then followed its mother down the path.

Chu Cuoi was amazed – was this a magic healing banyan tree? He wasn't sure he would know where to find it again, so he decided to pull it up and plant it at home so that he could help other people.

As he walked along the path with the tree balanced on his shoulder, he met an old man sitting on a tree, crying. 'What is the matter?' asked Chu Choi.

'I am at my end of days,' said the old man, 'my body is giving up, and I feel I have only days to live, yet I want to see my son and his children.'

Chu Cuoi thought and then said quietly, 'I have something that may help you – would you like to try?'

The old man grasped his hand. 'Please, just a few more days will help.'

Chu Cuoi took some of the leaves, chewed it in his mouth. It was very bitter. He placed it on the old man's lips. Within seconds the old man was drawing himself up, his pale face now had more colour to it. 'What is this gift that you have?' the old man asked.

Chu Cuoi explained. 'Then you must take great care of it,' pleaded the old man. 'Magic feeds from purity. Make sure that it is planted in pure soil, and you use only pure water. Care for it with a pure heart.' Chu Cuoi promised the old man he would do that.

He planted the tree in his garden. Soon he was known as someone who had a cure for people who doctors had written off as hopeless. His patients were generous with their thanks, and he grew wealthy. But each day he would water the tree with pure water, made sure the ground was kept clear, and he would meditate with pure devotion for the tree.

In time he married, and while at first his wife was pleased to help with the care of the tree, eventually she became jealous of the amount of time Chu Cuoi spent looking after the tree. She felt quite neglected. One day Chu Cuoi had to travel, and he asked his wife to look after the banyan tree for him. She agreed, but so resented the tree that she fed it foul water, threw rubbish all about it, then cursed it. To her surprise the ground rumbled, and the tree started shaking and pulling its roots from the dirt. Terrified, she watched as the tree broke loose from the earth and rose into the air.

Just at that point Chu Cuoi returned home and, seeing the healing tree rise, jumped to catch the low-hanging roots. The tree rose higher and higher, and Chu Cuoi look down at the ground rushing away from him, and his wife calling him to return. The tree went higher into the sky until it reached the moon, where it settled on the surface and sank its roots into the soil.

Chu Cuoi collapsed to the ground and fell asleep, shattered from his voyage. He was woken by a silver-haired maiden who said, 'I am the moon maiden. Welcome to your new home. No one ever leaves here.' He looked up into the sky and could see his world spinning there. He was never able to return to his old home. If you look at the moon, you will see the banyan tree and Chu Cuoi still staring at it in deep devotion.

> This story is told in Vietnam at the August festival of Tet Trung Thu –
> the moon festival. Children eat mooncakes, carry rice paper lanterns,
> and sing songs to Chu Cuoi, the man in the moon.

EUROPE

The Moon Tarrers (Estonia)

Once there was a very mean farmer who always tried to do things on the cheap. He wouldn't pay a penny if he could get it for a halfpenny! To him a day's work meant a day's work – and that meant that so long as you could see in front of your nose, then you could work. Especially when the moon was full in the sky and there was enough light to see your way.

Now, his two sons worked hard for their father, but they were exhausted at having to work during the nights when the moon was full. Their father even made them plough the fields in moonlight. They only got to sleep at night when the moon was in darkness. What should they do? If only every night was dark, and the moon didn't come out!

But what if the moon still came out but didn't shine so brightly! Maybe if they tarred it black, the moon wouldn't shine anymore, and they wouldn't have to work at night. What a brilliant idea! They got two very long ladders, a pot of tar each, and two good strong paint brushes. The next time the moon was full they got their ladders out, swung them back and forth until they were balanced on the moon, then swarmed up them to start tarring the moon.

The farmer woke up and couldn't hear them working – they usually made some noise. So, he got out of bed and looked through his window. He could see the two ladders reaching to the moon and two scallywags at the top, who he immediately recognised as his sons. Furious, he ran down the stairs, grabbed his coat and out the front door, over to the bottom of the ladders. He grabbed one and started shaking it. 'Come down here,' he said. 'You've got work to be doing!' Then he tried the other. 'I'm not paying you good money to paint the moon.' He rattled it again.

Now, what was a small shake down on the earth ended up being a huge shake in the sky – the ladders went flying back and forth, and the two sons tried to grab something to steady themselves. There was only the tarred moon. As the ladders started to fall to the earth, the two sons could only hang on to their tar brushes, now stuck firmly to the moon.

And if you look up at the moon when it is full, you can see the tar patches on the face of the moon – and maybe you can see the brothers too!

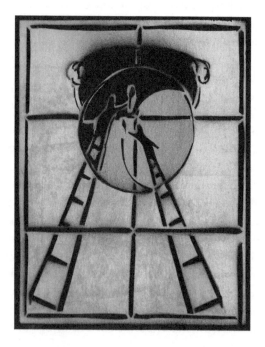

Daughter of the Moon, Son of the Sun (Siberia: Sami)

High in the sky, the sun makes his daily journey. It takes so long that in the dawn he has a polar bear pulling his golden sledge. By midday his sleigh is pulled by a buck reindeer, and at dusk by a doe reindeer. The sun is generous and kind. He grants life, nourishes the plants and trees, and gives strength to all living things.

He had a son, Peivalke, that he adored. As Peivalke came of age, the sun sat down with him, father to son. 'It is time for you to take a bride,' he asked. 'Have you considered whom you might marry?'

'Oh father,' replied Peivalke, 'indeed I have. Several earthly maidens seem suitable, but when I invite them to wear the golden slippers to ascend to the sky, they are too heavy and clumsy. I fear I may never find anyone.'

The sun was disappointed to hear this. 'Perhaps I will talk with the moon. I hear she has a baby daughter that in time might marry you. After all, they are poorer by comparison to us, and you, my son, are quite a catch.' The sun took his time until the day he could rise early and meet the moon in the sky.

'I have a proposition for you!' he called out. 'My son is looking for a bride, and your daughter might be the perfect match.'

The moon was horrified. 'My daughter is much too young for marriage. She is only a babe. Your son may scorch her, he can be so rough. When the time comes, I would like her to marry Nainas of the Northern Lights.'

'What does he have to offer?' roared the sun. 'I am the more powerful. I give life and strength to the whole world.'

'Yes,' replied the moon, trying to keep calm, 'but you are only alight for half the day. At night it is Nainas and his brothers that shine.'

The sun was indignant. 'My son will marry your daughter!' he declared and returned to his orbit. The moon sank down – her journey completed.

The sun was in a foul mood. The skies raged, with lightning crashing and thunder booming. The wind was torrential, tearing trees from their roots and houses from the ground. The waves on the sea towered above the cliffs. The reindeer huddled together, unsure what was happening but feeling that something was not right with the world. The Saami people trembled.

The moon knew she must protect her child. She scanned the earth, looking for somewhere safe. On a small island she saw a couple who yearned for a child. The moon placed her child in a cradle and left it close by. Their cries of delight when they saw the baby vindicated the moon, and she knew her child was in safe hands. She would be able to watch over her at night.

The couple brought the moon child up as their own, not concerned at her pallor. 'Everyone is equal and deserving of love,' they said. They were amused when the moon child stood outside the house at night and bathed in the rays of moonlight. They even referred to her as their moon child, which pleased the moon greatly. She was called Niekia. She grew slender and tall. Her foster mother taught her how to take reindeer skins and embroider them with beads and silver. The results were quite beautiful, and people travelled from afar to see and buy one of these skins.

The stories of such a maid reached the sun. He was still angry that the moon had managed to hide her child. So, he sent his son to investigate. As soon as Peivalke saw her, he fell in lust with her and desired her. He offered her the golden boots to try on, to become his wife.

She looked at him with distain. 'These would burn my feet!' she cried.

'Oh, I'm sure you would get used to it,' replied Peivalke as he placed them on her feet.

As soon as he did so, she turned into a misty haze. She wasn't the daughter of the moon for nothing! As the haze, she rose into the sky and hid amongst the moonbeams. Peivalke was horrified and thought he had killed her. Believing he had lost her, he returned to his father's house, convinced that he would never find a wife.

The moon delighted in having her daughter amongst her beams, but also knew that she was still not safe. The moon placed Niekia back on the earth and pointed to a pathway lit by the moonbeams.

'Follow this trail through the forest, over the plains, until you reach the ocean. There you will find a single hut. May you find your destiny there.'

Niekia did as she was instructed. The hut was bigger than she expected. She went up to the door – it was open. She called out, but no answer came. She stepped into the hut and was surprised that it was all one room. And very chaotic. With nothing else to do and nowhere to go, she started to tidy up. When it was spick and span, she found a corner of the room and turned herself into a spindle on the wall and waited.

Between the sun being full in the sky, and before the moon has risen, there is a point of twilight where the birds fall silent to mark the shift in worlds. A group of warriors came into the hut. Each was in silver armour, and each as handsome as the other. But the one that stood out was the last to enter. He was Nainas – eldest brother of the Northern Lights.

Nainas looked around the room. 'Someone has been kind enough to clear things for use. I feel that they are still here, although I cannot see them. Thank you for your good deed!'

The brothers sat down for their supper. When they were finished, they began their ritual battle dance of the Northern Lights. Their sabres clashed. Sparks flew that hovered in the air and became a curtain of light that ran out of the door across the descending darkness. Ribbons of white and red danced as the sabres struck each other.

Finally, the brothers laid down their sabres, and, exhausted, they sang songs about the glory days until one by one they left to return to their home.

Nainas waited until last. He called out, 'Spirit of the house. Thank you for what you have done for us. If you be of an age, then be as a mother to us. If you are of middle years, be as a sister to us. If you are young, then be a wife to me.'

Niekia was please that he spoke respectfully, and was generous in his gratitude. She revealed herself and he gasped. He
recognised her as the daughter of the moon. 'Will you let me be
your husband?'

She smiled, agreed, and they embraced. The first fingers of
dawn were crossing the sky. Nainas turned to Niekia. 'Tonight,
we shall celebrate – but for now, I must return to my home before
the sun rides his polar sledge.' With that he was gone.

Over the days it became a pattern of the twilight and the night.
The brothers would feast, then battle dance so that the Northern
Lights spread across the night sky. Afterwards they left one by
one. Nainas would pledge his honour to Niekia every night, then
leave her as the dawns rays poked through the darkness.

'Stay with me,' she pleaded, 'stay with me during the day so
that I have you close to me.'

He shook his head. 'I must return to my home; otherwise, if I
stay, the sun will pierce and burn me with his shafts of fire.'

This did not satisfy Niekia. An idea came to her. She found a
quilt of reindeer skin and embroidered on it the stars of the Milky
Way. When the brothers came, and were distracted by their mock
fighting, she placed it in the ceiling. After his brothers had left,
and the door was shut, as she lay in in Nainas' arms, it looked as
though the Milky Way was above them. When dawn began to
break, he opened his eyes and made to leave, but Niekia pointed
out the 'stars' above and he went back to sleep.

Eventually, Niekia awoke and climbed out of the bed. She
opened the door and went out into the full sunshine and felt its
warmth. But she had not closed the door behind her properly
and the sun's fingers sneaked into the hut. Nainas awoke and saw
the polar sledge in the sky. He tried to get away, but the sun saw
him and with no hesitation shot down a shaft of fire that went
through Nainas' body, trapping him to the ground. Niekia was
devastated. She tried to create shade for Nainas by standing over
him. He recovered enough to fly off to the safety of the skies.

The sun recognised Niekia straight away.

'You!' he cried. 'My son told me you were dead!' He grasped her by her long plait and held her close, burning her with his blazing stare. With one loud cry he called for his son – Peivalke.

'No!' gasped Niekia. 'You can burn me, roast me to a cinder. But I will never marry your son. '

The sun was furious and threw Niekia away into the heavens, where her mother the moon caught her in her arms, and held her tight to sooth her burning skin. If you look at the moon today, you will see the shadow of Niekia's face on her mother's breast. As the moon covers the night sky, Niekia looks down to see her beloved husband and his brothers battle dance in the night sky as the Northern Lights.

Sun Maiden and Crescent Moon (Siberia: Ket)

In a time gone by there were a girl and a boy. Their parents were killed when they were children, and they were brought up by distant family. Despite such a tragedy so young, they grew into strong, resilient young people.

When the boy became a young man he determined to travel, see the wonders of the world, and meet new people. He waved goodbye to his sister and began his travels. He revelled in the glory of every part of nature – from the soil and its teeming life under his feet, to the diversity of the plants, trees, life all around him, to the magnificence of the day and night skies that were above him.

In the sky, the Sun Maiden craved companionship. Looking down to the earth, she saw the young man with his daily prayers and heard him declare that he wanted to meet the Sun Maiden and travel with her. She was rather attracted by this thought and stretched down the long rays of her fingers to warm the earth around the young man. He was intrigued as the sun played around him and he reached out to embrace it. The Sun Maiden took that as consent and snatched the young man into the sky.

Folk Tales of the Cosmos

At first the young man was surprised, then very excited and interested. But too much can be a good thing, and he soon longed to return to earth, to recommence his travels and see his sister. 'Please let me go back,' he pleaded. 'I promise to return to you.'

The Sun Maiden was fearful. By taking the young man into her domain, he had now come to the attention of evil spirits that would try and harm him. Whilst he was with her, no harm would come to him. She tried to explain this – but the young man was determined.

'Very well,' she said. 'Take this whetstone and this comb to protect you. You will know what to do when the time comes.' He thanked her for them and put them in his knapsack.

'How do I get to the earth now?' The Sun Maiden smiled, and summoned her winged horse, who took them both down to the earth. The young man was pleased when his feet touched the cool soil of the earth close to the house of his sister. Tethering the horse, he ran joyfully to see her.

However, Hossiadam, a wicked sorceress, had arrived before him. She had killed the sister and eaten her. With her magic, she transformed herself into the shape of the sister, and it was into her arms that the young man threw himself in greeting. As the sister, Hossiadam offered to make a feast to celebrate his return. She placed a pot full of water on the fire to boil, and while the young man rested and went to sleep, she went out to the tethered winged horse. She cut off its leg and threw it in the pot to cook.

When the young man awoke, he chatted with his 'sister', telling her all about his travels. Then he saw what was cooking in the pot. It was the leg of his horse! He turned back to his 'sister' and stared at her. This time he began to see her for what she was: Hossiadam!

He pushed her over, grabbed the horse's leg from the cooking pot, and ran outside to find his three-legged horse standing there. He tried to replace the leg as best he could, untether the horse, then ride away. Hossiadam was behind him, screaming, shouting and cursing him.

The horse tried to rise into the air, but the damaged leg hindered him, and the young man tumbled to the ground. He rolled off the horse, laid his hand on its head to say, 'Thank you.' But he could not linger – Hossiadam was close behind. He ran.

Up in the sky, the Sun Maiden watched. There was nothing she could do.

The young man ran and ran. Hossiadam was always close behind him, spreading out her long fingers but not quite reaching him. Then he remembered the gifts of protection the Sun Maiden had given him. He thrust his hand into his bag, rummaged until he found the whetstone. Without a second look he threw it over his shoulder and kept on running.

Where the whetstone hit the ground, it began to grow. As Hossiadam approached, it was already a small hill. As Hossiadam got closer, it became a great mountain, blocking her way. The young man kept running, not looking back. Hossiadam was furious at being stopped. She tried to tear down the mountain, but to no avail. She bared her teeth and started to chew bits off. Slowly a hole appeared, and she managed to squeeze through to the other side. She could see the young man running in the far distance, and with a huge leap she caught up to him.

Gasping now, the young man felt the breath of Hossiadam on his neck. He rummaged again in his bag and found the comb. He threw it over his shoulder. The comb caught Hossiadam in her eye and, as she batted it away, the comb fell to the ground and turned into a great, dense forest. Hossiadam cursed again as she halted. This time she bared her teeth straight away, and quickly gnawed her way through the forest, creating a pathway.

The young man was tiring and Hossiadam caught up with him again. Her long spindly fingers reached out for him and just caught his left leg. In the sky, the Sun Maiden watched everything. She stretched her arm down to the earth, and with her fiery fingers caught hold of the young man's right leg. Hossiadam was on the ground pulling one leg, and the Sun Maiden was in the sky pulling the other leg. Both made one last grand effort to pull the young man. He was torn in two.

The Sun Maiden held her half in her arms and tried to fill him with her warmth. But this half did not contain his heart, and she could only keep him alive for a few minutes.

'I'm sorry,' she said. 'I cannot help you here. The only thing I can do is send you to the other side of the heavens, where you will live forever. We shall part but meet again on the longest day of the year.' With his last bit of life, the young man nodded and died.

With that, she used all her strength to throw him to the furthest part of the heavens. You can see him now as the crescent moon. With no heart, he appears cold and lifeless, but reflects the Sun Maiden's light.

The Woman Who Tarred the Moon (Sweden)

Atsisjoedne was a bad-tempered Sami woman who saw no good in anything and only wanted things for herself. She treated her own reindeer badly, even trying to beat them as she was milking them.

The reindeer had had enough, and decided they were going to break away from being mistreated and return to the wild. They tossed their reins into the wind and laughed when they saw them caught up in an old birch tree. Now the reindeer were free.

Atsisjoedne was furious to discover the reindeer had gone. She knew she would never be able to hunt down wild reindeer by herself, but she knew that a nearby encampment of Sami people had tame reindeer. She decided that she would steal them in the night, when no one could see her. However, the moon was shining so bright she decided that she would have to tar over the moon so that she would not be exposed.

She got a pail of pitch and started to tar the moon. Whether she wobbled and fell into the pitch or whether the pitch sucked her in, I do not know, but when you look at the moon at night you can see her there still, with the pitch and the pail.

A Garment for the Moon (Jewish)

The moon shivered. The sun shone warmly throughout the day, but the moon was always cold when she came out at night. Eventually, she took her complaint to the sun. 'It's not fair! I am always so cold,' she declared.

The sun grimaced. The moon had a point, and he didn't want to see anyone in discomfort.

'What you need,' he said, 'is a garment to keep you warm! I shall arrange it.'

The sun called all the clever tailors together and told them, 'Make a garment for the moon.'

The clever tailors all nodded their heads. 'Yes, of course.' But it wasn't so easy. The moon was a different shape every night! How could you take one set of measurements and expect to make a piece of clothing to change shape every night. It was impossible. The clever tailors would have to admit defeat. But who would tell the sun?

Now, if you have clever tailors, there must be simple ones. The people who get on with the everyday stuff and get overlooked when the fancy projects come. 'We can try!' they said. The clever tailors responded, 'If we can't do it, how can you?'

Yankel was one of the simple tailors. He had dreams, but never the opportunities. He had heard the story of a cloth that was made of light that could stretch to fit any size. He told the simple tailors, and they agreed – that sounded just what they needed! 'Please go and bring some back!'

He wasn't planning to be the hero and he certainly didn't know where to find the cloth. All he knew was the story. So, he packed himself some provisions and took his first step outside the city. Scared at first, he got bolder as he asked people he met if they had heard of a cloth made of light. He got used to hearing people laugh at him when they said, 'No! Such a thing cannot exist.' But he travelled on. Sometimes he had to sleep under a bush and felt the coldness of the night air. Sometimes he wondered if he was too foolish in his task, but then he would look up in the night sky and see the moon shivering. No one should shiver all the time.

He travelled many days until he came to a river. It was very wide, and he debated whether it was worth crossing it. Maybe he should give up and go home. Then, from the other shore, came a ferry boat. 'Do you want passage over the river?' asked the ferry man.

For one last time, Yankel asked his question. The ferry man laughed. 'Oh yes,' he said, 'I have heard of the cloth made from light, it's on the other side of this river. But only the queen can afford the cloth – the price for it is high!' Delighted that his goal seemed so close, Yankel paid the fare.

When he arrived in the city, he was surprised to see so many sad faces. 'It's the queen,' people said. 'She is sad because her daughter wishes to marry. However, tradition dictates that the queen can only wear her dress of the cloth of light to the ceremony. But the dress has started to unravel, and no one has the skills to repair it. No dress. No wedding. We are all sad.'

Yankel was surprised. Surely there were tailors in the city who could do the task. But they had all tried. 'It's impossible!' they said. He had heard that before.

He went to the palace and presented himself. The queen was delighted to hear that someone thought they could help. He was taken to her chambers and there she showed him the dress. He gasped. It was more beautiful than he had thought possible. It was very light, and so soft it floated in his hands. It shone with the splendour of the moon, and although he looked hard, he could not find the seams of the dress. When stitched, they had melted into each other. It was beyond magnificence, but as he explored the cloth, he realised that it would stretch and then return to its original shape. This truly would suit a garment for the moon.

He searched to find the unravelled edge. Indeed, a thread had come loose, and over time more and more had unravelled. Freed from the garment, the thread disappeared back into the light of day. Who had made this? How had they made it? The queen shook her head in sorrow. Generations ago the dress had been made for a past queen. It had come down to her, but now she was distraught because in her time it would eventually fade away.

'Let me study this,' he asked. The queen gave him permission to stay in her dressing room with the dress.

He held it up to the sun. He explored it every way possible. Pulling, stretching, seeing how it returned to shape. He could not see the seams by eye, but he knew where they must be. The sun went down, the moon rose. Even though it was a full moon, soon he would have to stop because there was not enough light.

Then, as the moon rose, the dress began to glow – it had its own pale light. Yankel held the dress up and to his amazement saw that the unravelled hem of the garment was growing in the moonlight, creating its own thread of light, and reweaving the cloth. This was the secret of the cloth that had eluded all others. It needed the full moon to fulfil its potential. Yankel knew now that this was what he needed. He took his scissors and snipped off a small section of cloth. In the moonlight that too grew.

When dawn came, Yankel took the dress to the queen. It was fully restored and there was no risk of it disappearing. She put it on, and the light of the dress reflected in her face and gave her a look of splendour. No wonder the dress had been handed down between queens.

She asked Yankel what he wanted for a reward. He showed her the small piece of cloth that he had snipped and asked for that. She laughed, told him to keep it, and gave him a bag of gold for his efforts. As he left the city, the bells were ringing to celebrate the forthcoming wedding.

Yankel had travelled far in his quest for the cloth of light. At each full moon on his journey, he brought out the snippet of cloth and let it grow. By the time he arrived at his own city, he had a substantial amount of cloth.

The simple tailors greeted him with joy. He had been away so long they thought he had been lost. Now they could show the clever tailors what they could achieve. It was with much pride that Yankel presented the sun with the garment for the moon that had been created by all the simple tailors working together.

The sun sent a message to the moon: he could finally keep his promise. She was overwhelmed with the gift and wore it straight away. She went from full to just a small crescent and back to full – but the dress perfectly fitted each time. She now rides in the night sky wearing the dress that was made from her own moonlight by the skills of the simple tailors.

NORTH AMERICA

Mouse-Woman and the Daughter of the Sun
(North-West Coast of America and Canada)

*'Mouse-Woman is the smallest of the Narnauks – supernatural
beings who inhabit the mythology of the Indians [sic]
of the Northwest coast of America and Canada …
Sometimes (she is) a tiny old granny and sometimes a mouse
with concern for young people … [She aims to restore] order
and balance.' Christie Harris, 1977*

There was a village in the north-west of the Americas, where the people lived in totem-pole houses that showed the emblems of the tribes: Eagle, Raven, Bear, Wolf, Frog and the Killer Whale.

In the darkness of winter, this village was covered in rainclouds over the sea and the forest. To protect themselves, the people wore robes of cedar bark. In the sky, the daughter of the sun was intrigued to know what it was like in the very dark patch of the north-west, hidden by rain clouds.

In the village, the young men and woman sported to find their partners. Suncloud was such a youth. His face was as dull as a raincloud hiding the sun, but his warmth and kindness shone through, and everyone loved him for it. Snowflower was a young woman who was very beautiful, but inside she was as cold as her namesake. People were attracted to her for her beauty but turned aside when she was cold and unkind to them.

Suncloud could not see the darkness in her. She was his cousin and they had grown up together, first playing, then fishing and sailing. He adored her and, in his heart, he wished to marry her.

But Snowflower was very vain. She did not want to marry Suncloud – she wanted someone handsome. 'Perhaps,' she thought, 'if people could see me being pursued by one man, then others would follow.' She would have her choice of suitors.

She told Suncloud that she could not marry him unless she had a white otter sea pelt. Suncloud knew they were very rare, and he set off to find one. He searched for a year but could not find the pelt. He sent up his prayers to the Great Sea Otter, who took pity on him and led Suncloud to a dying white sea otter.

When Suncloud presented it to Snowflower, she said, 'No, I will not marry you, unless we can have a goose feast in early summer when the roses bloom.' Suncloud was aghast – by the early summer all the geese would be in their summer camps in the far north. How could he achieve this? He asked around but no one knew. To manage his bleak thoughts, he went fishing and then gave his catch to an old woman who had no one to provide for her. She squeezed his hand in thanks, then told him that in the old days her father had spoken of a distant lake where some geese rested on their way north.

To prepare for his journey he stalked and killed a wild goose, then skinned, dried and stuffed it. He put it in his backpack, and after many days Suncloud found the lake. There were many geese – more than enough for Snowflower. He went to the lake and used the stuffed goose as a decoy, calling like a goose. Over several days the geese came to see the decoy and he was able to kill them. He carried them back to Snowflower and presented the bundle to her.

When she saw the pile of geese, Snowflower was horrified. She didn't want them! Was nothing going to put him off? But she was getting a lot of attention from the other young men, so she decided to string him along. She saw a slave walking by. The lowest of the low. Everyone instantly knew he was a slave because his hair was cut short – unlike the men and women of the tribe, who grew it long.

'Cut your hair,' said Snowflower. Suncloud was so impassioned he did it, even though it went against all his instincts.

'Now will you marry me?' he asked.

Snowflower laughed in his face. 'Marry you? A man who looks like a slave. Would I sink so low?' With that she caught the eye of another young man and walked away with him.

Suncloud could not believe how he had allowed himself to be manipulated by her. He was so ashamed, he ran away and hoped he would fade into the earth.

Mouse-Woman watched from her resting place. She felt sorry for Suncloud – he didn't deserve such treatment. But she was also concerned that others in the tribe had let it happen. Should she interfere? Where the happiness of one was destroyed in such a way, maybe yes?

So how could she teach a lesson and restore Suncloud?

Then it came to her. She knew the daughter of the sun wanted to visit the village hidden by the raincloud – maybe she could help to remind people that beauty is nothing unless there is warmth and kindness too. Time for a conversation.

Suncloud saw a tiny little mouse trying to climb over a log.

'Come, little sister,' he said, 'let me help you.' He scooped up the mouse and placed her on the other side. No sooner had the mouse scurried into the undergrowth than there was a crackle by his side. He turned, and there was an old woman – he was astonished, as he had not heard her on the path.

'You are a kind soul,' said the old woman. 'Come into my house.'

Suncloud was surprised to see a doorway wedged in the under-growth, and he had to duck as he followed the old woman in. He looked around, and the house was decorated with images and totems of mice. He suddenly realised that this was Mouse-Woman – the Narnauk – the carer of young people.

'I know of your trials,' she said, 'and I know of one who is more worthy of you. The daughter of the sun!'

'Why would she marry me? Surely, she would have other suitors?'

'I am sure she does, but none of them live in a raincloud village like your own. I know she would like to live there for a short time with humans.'

'Even for a short time, I would happily marry the daughter of the sun. How do I find her?'

Mouse-Woman smiled. 'Follow this track to the bottom of the mountain. Climb to the topmost part of the mountain, and there you will see a further track. Climb that until you reach the house of the sun. But whatever you do, do not look back behind you.'

Suncloud nodded. He found the track outside the house, fol-lowed it to the bottom of the mountain. Slowly he made his way to the top of the mountain, even though it was very hard under-foot. As he reached the top, a pathway opened in front of him, leading to the sun. He became a little afraid and wanted to look behind him to reassure himself that he was on a steady path, but then remembered Mouse-Woman's warning. 'Don't look back.' He took a step forward and breathed a sigh of relief that his foot was on solid ground. It was a long journey – his hair began to grow. He stopped to rest by the path, wondering what he was doing. What would the daughter of the sun be like? How would he know her? Perhaps he should return to his village and continue to pursue Snowflower. Then he remembered the Mouse-Woman's words. 'Don't look back.' He realised there was more than one way to look back. He shook himself – look forwards!

The path took him far into the sky and stopped outside a great house that shone gloriously bright. There was a man there. 'What do you want?'

'I have come to ask the daughter of the sun if she will marry me?'

The man laughed. 'Come with me,' he said.

He led Suncloud into the great hall. Everything shone beautifully, but not so bright as outside. The sun was at the centre of the room. His warmth filled every nook and cranny.

'What do you want?' the sun asked.

'I would like to marry your daughter. Mouse-Woman sent me.'

'Aha,' said the sun, 'then she must have good reason. There are several daughters of this house here – but you must identify which one is mine.'

He clicked his fingers and little sparks of flame fluttered. A woman stepped forward. Very beautiful, her face sparkled as though a light dust of frost rested, and there was something about her. Suncloud thought to himself, 'This cannot be the daughter of the sun; she is too cold.'

'Indeed,' said the sun, 'she is the daughter of the stars.'

He waved his hand, and the woman stepped back. Another stepped forward. She was equally as beautiful, but her face was as white as frozen cream.

'No,' said Suncloud, 'This is not your daughter. I would guess she is the daughter of the moon.'

'Well done,' said the sun. He gestured for a third to step forward. This time she was a beautiful as the others, but her face was radiant and warm.

'You are the daughter of the sun! Will you marry me?' cried Suncloud.

'Yes,' she said, 'I am willing to be your wife for a while. I long to visit your village under the raincloud, but I cannot stay too long. Are you willing to take me?'

'Oh yes!' said Suncloud.

With that the sun clapped his hands, a rainbow appeared, and the two of them slid back down to the earth. The villagers were overjoyed to see Suncloud return to the village with his new wife, Sunbeam.

Snowflower was furious – she was no longer the prettiest girl in the village. She tried to spread rumours about Sunbeam, but Suncloud's wife was so warm, kind and generous to everyone that no one believed the rumours. They realised that there was no value in beauty without warmth and kindness. Snowflower became so jealous that even her beauty brought her no friends, and people began to realise her nastiness. All her suitors turned away from her, and soon there was only one man left who wanted to marry her. Frightened of being alone, she married him.

Mouse-Woman smiled. It had been a bit naughty interfering like that, but it had all worked out. The daughter of the sun found out what it was like in a raincloud village, Suncloud had restored his belief in himself, and Snowflower knew what jealousy was like. The villagers had all seen that beauty is more than skin deep.

Oceania

The Blue Fish and the Moon (Australia)

In the Dreamtime, when all was still being made, and everything was new, there was no moon or stars in the night sky – everything was in darkness. In the daytime there was the sun.

There were two men. One was Nullandi – he was known as the happy man. He would find good in everything, even when it was a challenging time. The other was Loolo. He was known as the miserable man – no matter how good things were, he would always find something to complain about. They both lived with their wives.

In the darkness of the night, Nullandi would give thanks that his eyes were rested from the constant glare of the bright sun, and that his body was cool from the heat. His wife said that there was nothing to see in the dark. He laughed and asked her to look in the flames of the fire then into the darkness, and finally look up into the sky. She could see the thin pricks of light that were the stars.

'They are the sparks of fire that Baiame puts in the sky, but it is a mystery to us why he does it. You will only know this when you are dead,' said Nullandi. 'I shall be like Baiame, the creator god.' He took a burning branch from the fire and waved it around, the sparks flying into the air. 'See, Baiame's light is always there waiting for the dawn when we can see its yellow glow, but the stars in the night reflect the flame and promise of Baiame of a new day. When we die, our spirits will be like the sparks that we follow to Bullima the great sky camp, the spirit land, and discover the mysteries that only the dead can learn.' His wife laughed at the adventure. Then he put the branch back on the fire. 'I should not have done that. No man can be like Baiame. Sleep now until the gift of a new day.'

For Loolo, the darkness brought fear and trepidation. His wife listened in the dark to his words and grew very afraid. 'When we die, we shall be nothing but dust that the wind scatters. We will be wiped from the face of the earth.' His wife sobbed at the futility of the life they had.

When Nullandi and Loolo met, Loolo asked why Nullandi's wife laughed at night. In turn, Nullandi asked why Loolo's wife cried.

Loolo replied, 'I tell her that death is like the darkness that falls at night. There is nothing. We fall away to nothing. There is nothing to see.'

Nullandi replied, 'Death is for a short time; the great spirit did not create us in this beautiful world just to bloom and fade away to nothing. He takes us to an even better place to live.'

Loolo scoffed. 'This world leaves us hungry, thirsty, full of fear and pain. There is nothing else.'

Nullandi shook his head. 'If you want to be a blue fish in the sea when you die, then be a blue fish. But at least let your wife be happy in this world.'

Loolo shook his head. 'You should let your wife make up her own mind, instead of filling her head with fanciful stories! What will she do when your body is ravaged by dingoes?'

Nullandi reached out to Loolo. 'We shall leave it in the hands of Baiame. He will decide what is to be the fate of everyone. If there are people like you who are afraid of the dark, then so be it. May Baiame make me become the light to comfort them.'

With that the two men parted, never to meet again.

Eventually, Loolo died and became a blue fish, hiding in the bottom of the lake until he was eaten by a bigger fish with his bones scattered on the seabed.

When Nullandi died his spirit was taken into the heavens by Baiame, to become the moon to bring light in the darkness for all living things. He waxes and wanes to remind mankind that even when things seem lost and darkness falls, there is always another time when the moon is full and shares its light, just as man's spirit dies for a while and then grows back.

And that is how Nullandi, the happy man, became Bahloo, the happy god of the moon.

Rona: The Woman in the Moon (New Zealand: Māori)

Rona was a woman who knew her own mind, and she expressed it with a quick temper and a sharpness of tongue. Sometimes the words cut deep in the hearts of those around her.

Her husband was a patient man, and often away for a night or two to fish. The best time was when the moon was full, so he said his farewells, and went with his fishing tools. The next evening, in anticipation of her husband's catch, she set up the cooking fire and heated the stones about it. She realised she had no water to sprinkle on the stones to create the steam, so she took a gourd and went down to the spring.

The moon was bright and shining, and she could see the path as if it was day. But then a cloud obscured the moon, and everything became dark. Rona could not see her way. In the darkness she fell over a ngaio tree root and stubbed her toe. The pain was so sharp – she cursed the tree, she cursed the cloud, but most loud of all she cursed the moon for the lack of light.

Marama, the moon, heard and was not pleased. The insults to him came thick and fast, so that to silence her he came down and tried to snatch Rona away. In a panic, Rona held on to the ngaio tree, but as Marama pulled her away, the tree came too and they were both taken up into the moon, where you can see their shadows today.

When her husband came home, all he found was the hole where the ngaio tree had been. He never guessed his wife was in the moon above him.

Maaui Tames the Sun (New Zealand: Māori)

Maaui is one of the demigods of the peoples of Polynesia and New Zealand. There are many stories of his birth and his deeds too numerous to tell in this moment, but there was one act he undertook that affected the lives of all those on this earth.

Tee Raa, the sun, brought light and warmth to the earth – but he passed over the sky so quickly that there was hardly enough time to even hunt, let alone cook and prepare food before the darkness of the long night fell about everyone.

Maaui could see that if they had enough time to enjoy the day, everyone would benefit and have time to enjoy the pleasures it might bring. He determined to slow down the sun. His brothers laughed at him.

'How are you going to do that?' they cried.

But Maaui smiled and nodded. 'I have a plan! Will you help me?'

The brothers were not convinced. 'We'll get burned!' they said.

Maaui threw up his hands. 'I promise you, these are the only hands that will burn! Now, will you help me? It will be much easier if we all work together.' The brothers agreed.

Maaui had had a vision. In it he had seen how to make flax, how it could be twisted and plaited to make very long ropes. He taught his brothers how to do this. They were surprised at the length they were able to get from such short pieces of flax. Maybe there was a chance that Maaui's plan would work.

When the ropes were finished, they began the long journey to the Cave of the Sun – Te Rua-o-te-raa. The brothers positioned themselves around the horizon, and as the sun rose, they slipped the ropes over it, criss-crossing until the sun was held tight and it strained against the ropes.

'Let me go,' called the sun, 'you do not know the consequences of what you are doing.'

Maaui called back, 'Then slow down as you cross the sky. You do not understand the consequences of your speed.' The sun just growled.

Maaui then pulled out his weapon, the jawbone of his ancestors Muri-ranga-whenua, and began to beat the sun! The sun tried to get away, but the brothers took the strain on the rope.

'Why do you beat me so?' cried the sun. 'I am Tamanuite raa – the sun. Show me respect.'

'We respect you in many ways, and want more time for the earth to breathe in your light and warmth,' responded Maaui. 'For men, women and all creatures to have more time to spend their lives in your light. We want you to slow down in your path. I will not harm you, if you agree to go slower.'

'I am who I am, and I travel the path that is set for me,' retorted the sun.

Then Maaui raised his weapon again and again until the sun cried out, 'Enough – I will do as you asked.'

Now the sun goes slower across the sky and there is time for all creatures to hunt, to cook, and to take pleasure in the day. Some say the sun is still limping because of the blows Maaui gave him, but from his wounds and scars come the sunbeams that dance in our fingers.

How to Steal the Moon (Caroline Islands)

The chief of the Caroline Islands had a beautiful daughter who came of age to marry. But no matter where he looked, he could not find a man worthy of being her husband or whom his daughter might possibly want to marry.

What to do? He went to his favourite place to think, and to look at the moon. She always inspired him with her grace and the gentle shine of her face. Then it came to him. If someone was clever enough to bring him the moon from the sky, then, surely, they would be worthy of marrying his daughter. So, the word was spread near and far. Many came but none succeeded.

There was a man from a poor family, who did not have many advantages in life, but he trusted in the wisdom and guidance of his mother. 'How can I steal the moon?' he asked. She smiled and told him of an old story of an invisible path that would lead to the moon.

'Go down to the beach and find the bent palm tree. When the moon is in the sky, stand between it and the tree. Raise up your arms to the moon and let the power of the wind raise you up to the path. It will shine before you and lead you to the moon.'

Thanking his mother, he did as she advised at the next full moon. There indeed was the path, and he stepped out on his way to the moon. It was a long journey, and to his astonishment he met people on the path. He called out to them. They smiled back and without a word gave him two plovers, two roosters, one pandunus fruit and a hibiscus stick.

When he came to the end of the path, he was outside a house of the gods. This was a place where Yalulep, the Carolinian high god, resided. No earth-born man or woman was allowed to enter. Inside he could see the moon, hanging from the rafters – the moon that belonged to Yalulep.

The young man began to tell stories to the guards; they became so enraptured they began to fall asleep. Leaving his bag of gifts outside, the young man made his way into the house. Everyone else was asleep. There, in the rafters, the moon was shining. He climbed up and untied the moon. It was so bright he placed it inside his clothes so that no one could see it, and crept out of the house.

He quickly found the path and his bag with all the gifts from the road. Then he started running!

It didn't take Yalulep long to realise that his beloved moon was gone!

'Thief! Thief!' he called and sent one of his guards to get back his moon.

The young man pulled out the plovers from the bag and threw them onto the road. They began fighting. The guard was entranced and stopped to watch them, then shook himself and chased after the young man again. The guard was so close to him that he could almost catch his ankles, but the young man reached into his bag and pulled out the two roosters. They fell against the guard and started to attack him, trying to peck out his eyes. Again, the guard shook himself and the roosters fell to the ground and started to fight each other. The young man ran, and again the guard caught him up. This time the pandanus fruit was thrown to the ground, and a spiky bush sprang up. The guard was held back until he managed to crawl through the bottom of the bush near the roots. He was bleeding, but that still did not stop him chasing as fast as he could.

The end of the path was in sight, the young man could hear the guard coming closer to him. There was only the hibiscus stick left in the bag. He threw it behind him. A great thorny bush grew up across the path. This time the guard could not get through.

As the young man reached the beach, he stepped off the path, and turned to look at it. The path was gone. But he had the moon! He kept it tight to his chest so that no part of the light shone out, and went straight to the chief.

'Here is the moon,' he said. 'Keep it covered, otherwise it will get stolen away.'

The chief could see just a tiny little light under the covers. He was thrilled that he now had the moon for himself. 'You shall marry my daughter!'

There was a grand wedding. In the evening there was no moon in the sky. People commented on it. Should they be worried? The chief smiled to himself. The moon belonged to him. He crept back to his hut and took the moon out of the basket where he had hidden it.

'I'll just peek,' he said. He took one of the covers off. The light shone a little brighter from under the rest of the covers. 'Just another cover,' he said. 'Nothing much is happening. It will be quite safe.'

Slowly he took all the covers off, and there was the moon in all its glory. But no sooner had the last cover dropped to the floor, than the moon flew from his hands, dodged around the room, and then out into the world, back into the sky.

The chief was furious. He summoned his new son-in-law and demanded that he bring back the moon. The young man shook his head. 'This way,' he said, 'when the moon hangs in the sky, everyone will know it really belongs to you, rather than Yalulep.'

The chief thought about it and rather liked the idea of that!

The young man and the young woman then got busy being husband and wife under the full moon.

SOUTH AMERICA

The King and the Moon (Dominica)

There was once a king who wanted everything he could see. He had a palace full of clocks, hatstands, china bowls, bicycles, so many books and much more. But he didn't use any of them. He just liked to look at them all and know that they belonged to him. They sat there gathering dust, so he collected servants just to clean them. The palace was bulging with things that he didn't need. He was even thinking of collecting palaces so that he would have plenty of room to put even more things!

One night he was admiring some of his hoard, when he chanced to look out of the window. The moon was very full that night. It had an aura around it that made it particularly beautiful. There were no clouds in the sky, and it just hung there, with the stars twinkling behind it.

The king looked at it in amazement. He knew that there was a moon, of course, but he had always been so taken with his huge collection of things that he had never looked outside his palace. Now he saw it he knew he must have it!

'Bring me that moon!' he said.

His servants looked at each other in despair.

'Bring me that moon!' cried the king, again.

His servants shook their heads. They had no idea what to do.

'Bring me that moon!' roared the king.

The officers of the court shook their heads. The chancellor stepped forward.

'Sire,' he whispered, for fear of upsetting the king. 'Sire, it is too far away to reach.'

'Nonsense,' said the king, 'all it needs is a big tower. Build me a tower out of boxes.'

The servants weren't sure what to do at first, then one dragged a box out, then another.

'That's it,' said the king, who was now getting very excited. 'More boxes.'

The tower grew higher and higher.

'More boxes!'

The king couldn't wait. He started to climb the tower of boxes. 'I'll get it myself.'

He got to the top. The moon seemed just out of his reach.

'More boxes.'

They had used all the boxes in the palace, in the kingdom, and were now bringing in boxes from all over the world.

The king had never had so much fun. He stood at the top of the tower, stretching for the moon.

'Just one more box will do it!' said the king.

But there were no more boxes left in the world. Someone shouted that up to the king.

'I need one more box,' he called back, 'then I can reach the moon. Just take the bottom box and pass it up!'

The servants and the officers of the shook their heads and widened their eyes in disbelief at what their ears heard. The bottom box? Pass it up? How would that work?

'Come on,' said the king, 'I haven't got all night. Pass up the bottom box!'

Very carefully, the servants gave the bottom box a tug. Slowly, slowly they eased it out. They were just halfway there when the tower began to rock.

'Hold on Your Majesty!' called out the chancellor.

The tower went this way, then that way! Then there was a moment when it seemed the tower was a snake curving into the sky, then *whoosh*!

All the boxes came tumbling down. The king, at the top, made one last attempt and flung himself into the night sky, hoping to grasp the moon! And that is where you can still see him today.

So, beware what you wish for!

The Fifth Sun (Aztec)

When the Aztec gods first created the earth, there were just little pockets of light in small crevices on the surface. It was too dark to properly see the new creation, so the gods collected all the bits of light and created a big ball of light to be a sun. Who should have the honour to carry it up into the sky?

Before anyone could say anything, Tezca, one of the minor gods, grabbed the sun, put it on his back, and leapt into the sky. 'Oh,' said the other gods, 'he is there now, let him get on with it.'

Unfortunately, Tezca was not strong enough to shine evenly – spots of light would sit next to spots of shadow, and often by midday, he was too tired to shine. The gods told him to come down, but he wouldn't, so Quetzalcoatl, the plumed serpent, took a big stick and knocked him out of the sky. Tezca was ashamed.

The gods then made another sun, but it was too dull to shine brightly. The third sun was too weak, and a windstorm blew it away. The fourth sun was the wife of the rain god, and as well as shining light, it rained all the time and there was a great flood.

The gods planned a fifth sun. They sat around their fire and wondered who would carry the sun this time, who would shine brightly to light the world – who would sacrifice themselves for this deed. Because this sun needed to shine brightest of all, and for the rebirth of the sun there would need to be a death of one of the gods.

'Me!' said Tezca. 'Give me another go.' He wanted to show that he could do better than the others and wanted to regain some of his past glory. Some of the gods nodded in agreement, but some of the other gods looked around for someone else.

'You!' said one of the gods, pointing at Nanahuatzin, who sat in the shadows to hide – old, twisted, covered in scars and scabs. Other gods nodded in agreement. 'Me?' said Nanahuatzin, astonished. 'You don't want me!' But the gods persuaded him he should try, and Nanahuatzin agreed.

It was decided that the two candidates would each have a hill to prepare for the ceremony, to fast and to purify themselves. This was in a place known as Teotihuacan, and even today the hills are known as the Pyramids of the Sun and Moon.

The gods built a bonfire that burned for four days, each day getting even hotter. Meanwhile, the candidates prepared themselves for sacrifice by bloodletting and making the prescribed offerings of fir bough, bundles of bound grass and manguay thorns dipped in their own blood, and incense.

Tezca was determined to be remembered for the exquisiteness of his offerings. He used brightly coloured and much prized quetzal feathers instead of the fir bough, balls of gold instead of the grass, spikes of jade topped with coral instead of thorns with blood, and the finest, most costly incense.

Nanahuatzin had little to offer but fir boughs and the manguey thorns with his own blood. Lacking any incense, he could only pick the scabs from his scars to make their own fragrance.

As the fourth day dawned, both men were brought to the bonfire. Tezca was radiant in the most magnificent clothes, while Nanahuatzin wore only paper. The gods had been up all night drumming a rhythm, so their hearts beat as one. The two candidates of such contrast stood before the raging pyre, their faces flushed and sweating.

'Who will jump first?' cried the gods.

Tezca turned to the gods. 'I will be bigger and better than any of the four suns you have had before. I know the burden, and now I have the strength to shine all day all over the earth.' With that, he turned back to the fire and ran towards it. The gods held their breath. But at the last moment Tezca stopped and turned away. He waved to the gods as if this was all his plan, then walked back to run again. Three times he reached the blazing fire, three times the heat overwhelmed him, and his courage fell away. The drumming continued. For one last final time Tezca ran towards the fire, then found himself frozen in the heat, shamed at his cowardness and unable to move. The drumming stopped. Everyone watched to see what would happen next.

Nanahuatzin, with no words nor gesture, took a step forward, then walked slowly and deliberately towards the pyre. He passed Tezca, brushing his arm, then with one final cry he hurled himself onto the fire. Rays of light emitted from the fire, filling the sky and the earth for a few seconds, and then there was darkness, with only the light from the fire.

Tezca shook himself, knowing that if he turned back now, he would be shamed forever. He took a step forward and then flung himself onto the fire.

In the east the sky reddened, then a great shining light that covered the known earth rose. It was so bright that the gods had to shield their eyes. This was Nanahuatzin, who had been transformed into Tonatiuh, the fifth sun. The heat and light radiated from him, and the gods knew that this was the right thing.

Someone pointed at the horizon. Another sun was rising, just as bright – it was Tezca, trying to outshine his rival for glory. Two suns were too bright and too hot. The gods shouted at Tezca to dim himself down, but he ignored them. Then, in a moment of panic, one of the gods grabbed a rabbit that was watching and hurled it into the sky towards Tezca's sun. The rabbit's impact obscured some of the brightness and heat, and you can see the imprint of the rabbit on the moon today.

Thus, the fifth sun was created with the moon beside it. But they both just hung in the sky, not moving. The gods sent a falcon to ask why they did not cross the sky. The reply came back that a blood sacrifice from all the gods would be needed.

The gods were aghast that this should be asked of them, but there was no choice. They looked about the earth they had made, saw that it was good, and prepared themselves for sacrifice. They took off their fine clothes and laid them on the ground. In time the clothes would become the relics by which they were worshipped. Then they pierced their hearts and fell to the ground. Satisfied, Tonatiuh the Sun moved in the sky, with the moon waiting a respectful distance before it followed.

In this way came the sun of the fifth age in which we live.

ACKNOWLEDGEMENTS

Many and various people have assisted with this collection of tales. Some people provided the initial story from which I was able to find a written or online version. They are acknowledged by story contributed.

Many and various people have helped me with trying to work through the issues of cultural appropriation. The tradition bearers know who they are, and I respect your wish for anonymity as none of you want to be seen as the sole spokesperson from your community. I thank Liz Weir, Debbie Guneratne and Fiona Collins – members of Equity Union who produced the Equity Guidance Notes for Storytellers on Cultural Appropriation. Individually they spent time with me to discuss my approach to collecting the stories compared to the aims of the guidance. Whilst my interpretation of the guidance notes was informed by these discussions, I take personal responsibility for the way I have applied them. If I have caused any offence or been insensitive, I apologise and please let me know so that I can change future editions.

Many thanks to the members of the Folk Tales of the Cosmos Facebook page, who debated whether I should organise the stories by culture or by the Greek name for the constellations. They were evenly split between both. In the end I decided to go by continent, as the group were most likely to be seeing the same piece of night sky – although recognising that places like Africa stretch from north of the Equator to just above Antarctica, and within each country making up Africa there may be several cultures each with their own stories.

Many thanks to the people I met in the online storytelling groups during lockdown, who were willing to share material or at least give me sufficient information to track down the stories. It's much easier to research something that you know exists than to randomly search.

Many thanks to the members of the writing group I attend for their useful comments and feedback. As non-oral storytellers, it was useful to get their perspective on the stories.

Many thanks to the members of the Sid Vale Storytellers Group, who were patient with me as I stumbled through some of the stories as a first cut at telling them.

Many thanks to Nicola Guy from The History Press, who saw this project grow from the planned two to five years, was patient with me during the two years of lockdown (as many creatives struggled), and was supportive when physical health problems involved operations and delayed the project even more. Many thanks for support from Debbie Pearce during the very dark days.

Many thanks to Vicky Jocher, who valiantly offered to do the illustrations for the rest of the world tales and made such a splendid job with them.

Many thanks to Chris Dowling, Pam Dowling and Roisin Murray for help with proof reading.

And not least – many thanks to Jeff Ridge, my partner in life, who prepared the illustrations for the Greek tales, which are from Alexander Jamieson's *A Celestial Atlas*, published in 1824.

Notes on Individual Stories and Sources

Part One: The Greek Myths of the Stars

I have known the Greek myths all my life, so it's hard to pick on sources. These are the ones on my bookshelf that I used to check things.

Falkener, D.E., *The Mythology of the Night Sky* (Springer, 2011)
Galat, J.M., *Dot to Dot in the Sky: Stories in the Stars* (Whitecap Books, 2001)
Griffiths, A.M.M., *The Stars and Their Stories: A Book for Young People* (1913)
Moore, P., *Legends of the Stars* (History Press, 2009)
Williams, M., *Greek Myths for Young Children* (1991)
Other books in the Bibliography.

Part Two: The Rest of the World: Stars

Africa

The Rabbit Prince (South Africa: Shangaan)

Alcock, P.G., *Venus Rising: South African Astronomical Beliefs, Customs and Observations* (2014) p.306; assa.saao.ac.za/wp-content/uploads/sites/23/2014/08/Venus-Rising-2014-P-G-Alcock.pdf (*This book gives a comprehensive overview of South African Astronomical beliefs, etc. It covers various cultural groups of South Africa in detail and is a must read for an in-depth understanding of the cultural contexts.*)

Bourhill, Drake, *Fairy Tales from South Africa* (1908) p.43; archive.org/details/fairytalesfroms00hollgoog/page/n68/mode/2up

The Warrior, the Khunuseti and the Three Zebras (Bushmen: Namaqua)

Alcock, P.G., *Venus Rising: South African Astronomical Beliefs, Customs and Observations* (2014); assa.saao.ac.za/wp-content/uploads/sites/23/2014/08/Venus-Rising-2014-P-G-Alcock.pdf

The Stars and the Star Road (Africa)

This story appears in many places – but I used this one.
www.worldoftales.com/African_folktales/African_Folktale_17.html

Asia

The Cowherd and the Weaving Girl (China)

Martens, F.H., *Chinese Fairy Tales* (1998)
Roberts, M., *Chinese Fairy Tales and Fantasies* (Pantheon Library, 1984)

Krittika: The Seven Wives of the Rishis (Hindu)

Many thanks to Dr Swati Gola for assistance.
Adama, M., *Seven Sisters of the Pleiades*
Ganguli, K.M. (trans.), *The Mahabharata of Krishna-Dwaipayana Vyasa* (1896) www.sacred-texts.com/hin/maha/index.htm
John, C.P., 'The Big Dipper in Ancient Indian Astronomy' (2022) grahamhancock.com/johnc1
Pattanaik, D., 'A cluster of goddesses' (2019) www.mid-day.com/news/opinion/article/a-cluster-of-goddesses-21814163

Europe

The Eyes of Thiazi (Norse)

Crossley-Holland, K., *Norse Myths* (1993) p.30
Gaiman, N., *Norse Mythology*, p.163
Guerber, H.A., *The Norse Men* (Senate, 1994) p.103
Lancelyn Green, R., Myths of the Norsemen (1960) p.46

The Silver Woman (Siberia: Sami)

Riordan, J., *The Sun Maiden and the Crescent Moon* (InterLink, 1991) p.92

The Veil of Imatutan (Estonia)

This one I was told by Gudrun Rathke, who had heard it from a German storyteller who said it was Latvian. I have subsequently found other versions of it

in Estonia, Finland and Russia. This is a composite of those versions.
Judd, M.C., *Project Gutenberg Classic Myths Retold* (1901; where it is called
   'The Milky Way'.)

The Seven Stars (Armenia)

Marshal, B.C. (trans.), *The Flower of Paradise and other Armenian Tales*
   (Libraries Unlimited, Detroit Wayne State University Press, 2007)

North America

The Great Bear and the Six Hunters (Seneca)
There are many indigenous stories that link bears to the Plough/Big Dipper.
www.sacred-texts.com/nam/iro/sim/sim87.htm

Follow the Drinking Gourd: The North Star to Freedom
www.nps.gov/articles/drinkinggourd.htm
Article by Julie West, Communications Specialist, NPS Natural Sounds
   and Night Skies Division
www.awesomestories.com/pdf/make/133709
Underground Railroad – Slaves Follow the Big Dipper
www.followthedrinkinggourd.org/What_The_Lyrics_Mean.htm
(This website reveals the code/riddle words that describe the route to be
   taken north.)

Seven Wise Men (Lenape/Delaware)
Hitakonanu'laxk, *The Grandfathers Speak: Native American folk tales of the
   Lenape People* (Interlink Books, 2005) p.86

The Children of the Northern Lights (Arctic and Greenland)
Millman, L.A., *A Kayak Full of Ghosts: Eskimo Folk Tales* (Interlink Books,
   2004) p.49

How Coyote Scattered the Stars (Navajo)
Monroe, J.G. & Wlliamson, R.A., *They Dance in the Sky: Native American
   Star Myths* (Houghton Mifflin, 1987)

Origin of the Pleiades (Onondaga)
*The Onondaga people are one of the original five constituent nations of the
   Iroquois (Haudenosaunee) Confederacy in north-east North America.*

Also see www.onondaganation.org/aboutus/
www.firstpeople.us/FP-Html-Legends/OriginofthePleiades-Onondaga.html

Oceania

The Husband and Wives Who Became Stars (Australia)
A.W. Reed, *Aboriginal Stories* (Reed New Holland, 1994) p.60

The Three Brothers (Australia: Yolngu)
Norris, R. & Norris, C., *Emu Dreaming: an Introduction to Australian Aboriginal Astronomy* (2009)

The Emu in the Sky (Australia: Northern Territory)
This is a different kind of star story in that it is the gap between the stars that is identified as a creature. It can be seen in the Milky Way.

www.abc.net.au/science/articles/2009/07/27/2632463.htm Australian Broadcasting Corporation

Irdibilyi, Wommainya and Karder (Australia: Torres Strait)
Wilson, B.K., *Tales Told to Kabbarli: Aboriginal Legends Collected by Daisy Bates* (Angus & Robertson, 1972)

The Trials of the Girls (Australia)
Smith, W.R., *Myths and Legends of the Aborigine* (Senate, 1930) pp.345–50
Reed, A.W., *Aboriginal Stories* (Reed New Holland, 1996) pp.77–81

South America

The Path to Abundance (Argentina: Toba Indians)
Stryer, A.S., *The Celestial River: Creation Tales of the Milky Way* (August House, 1998)
Metraux, A., *Myths of the Toba and Pilaga Indians of the Gran Chaco* (1946) p.20
at babel.hathitrust.org/cgi/pt?id=mdp.39015010561051&view=1up&seq=43

The Llama Star (Peru: Incas)
www.learnthesky.com/blog/star-stories-the-llama-star-an-inca-tale-about-the-southern-cross

Ganeri, A., *Star Stories: Constellation Tales From Around the World* (Templar Books, 2018)

PART THREE:
REST OF THE WORLD: MOON AND SUN

Africa

A Home for the Sun and Moon (Nigeria)

Kantor, S. (ed.), *101 African American Read Aloud Stories* (Black Dog & Leventhal Publishers, 1998)

Pritcher., D., *The Calabash Child: African Folk Tales* (David Philip, 1988)

www.sacred-texts.com/afr/fssn/fsn18.htm lk Stories From Southern Nigeria, by Elphinstone Dayrell, (1910), at sacred-texts.com

Asia

The Buddha and the Hare (Sri Lanka)

Richard Marsh, personal communication

Black, W.G., 'The Jataka Tale of the Selfless Hare'

Black, W.G., 'The Hare in Folk-lore', *The Folk-Lore Journal*, Volume 1 (1883)

Shedlock, M.L., *Eastern Stories and Legends* (1920)

Chang E: Goddess of the Moon (China)

Chang E is the goddess of the moon – her story is told during the mid-autumn festival on the fifteenth day of the eighth lunar month (September or October) when special pastries are put out for her to bless them and she bestows beauty on her followers.

Although the key characters and actions points are common, there is a lot of variation in how they came to pass. I have retold the tale drawn from multiple sources. I have tried to remain faithful to the spirit of the story.

Martens, F.H., *Chinese Fairy Tales* (1998)

Carpenter, F., *Tales of a Chinese Grandmother* (1973)

Chu Cuoi: The Man in the Moon (Vietnam)

De las Casas, D., *Kaminshibhai Story Theatre* (Libraries Unlimited, 2006)
The Moon Boy, a legend from Vietnam
www.vietnamtourism.org.vn/attractions/culture/myths-legends-folklores/
    the-moon-boy-a-legend-from-vietnam.html

Europe

The Moon Tarrers (Estonia)

Smith, Ure, *Australian Children of the World* (1979) p.77

Daughter of the Moon, Son of the Sun (Sami: Siberia)

Riordan, J., *The Sun Maiden and the Crescent Moon: Siberian Folk Tales*
    (Interlink Books, 1991) p.197

Sun Maiden and Crescent Moon (Siberia: Ket)

Riordan, J., *The Sun Maiden and the Crescent Moon: Siberian Folk Tales*
    (Interlink Books, 1991) p.55

The Woman Who Tarred the Moon (Sweden)

Hatt, E.D. (trans. Sjoholm, B.), *By the Fire: Sami Folktales and Legends*
    (University of Minnesota Press, 2019) p.52

A Garment for the Moon (Jewish)

*I first heard this Jewish story from Shonaleigh Cumbers as part of The Drutsyla
oral tradition passed down by Edith Marks. It appears in in the Gem Subcycle
'The Ruby Tree', also the 'Cloth of Hope and Sorrow' from the Menasseh Cycle of
Stories (c.1600). As this is a living tradition, Shonaleigh requested that people
do not tell from her version but find another source. I found a written source in
this book and based my retelling on that.*

Schwatz, H., *Miriam's Tambourine: Jewish Folktales from Around the World*
    (OUP, 1984)
Shonaleigh has now published her version.
Cumbers, S., *A Garment for the Moon* (Orkneyology Press, 2024)

North America

Mouse-Woman and the Daughter of the Sun
(North-West Coast of America and Canada)

Harris, C., *Mouse-Woman and the Mischief Makers* (Macmillan, 1977)
pp.91–105

Oceania

The Blue Fish and the Moon (Australia)

Reed, A.W., *Aboriginal Stories* (Reed New Holland, 1994) p.49

Rona: The Woman in the Moon (New Zealand, Māori)

Reed, A.W., *Māori Myth and Legend,* p.97

Grace, Wiremu, *Rona and the Moon Adaptation*
www.careers.govt.nz/resources/tools-and-activities/the-magic-of-myths/
rona-and-the-moonko-rona-me-te-marama

Best, Elsdon, *Maori Religion and Mythology Part 2: Lunar Myths* nzetc.
victoria.ac.nz/tm/scholarly/tei-Bes02Reli-t1-body-d4-d5-d3.html

Maaui Tames the Sun (New Zealand, Māori)

Reed, A.W., *Maori Myth and Legend,* (1983)

How to Steal the Moon (Caroline Islands)

Flood, B. et al, *Pacific Islands Legends* (The Bess Press, 1999)

South America

The King and the Moon (Dominica)

theracetoread.wordpress.com/tag/the-King-who-wanted-to-touch-
the-moon

The Fifth Sun (Aztec)

Bierhorst, J., *The Hungry Woman: Myths and Legends of the Aztecs* (Quill, 1984)

Ferguson, D., *Tales of the Plumed Serpent: Aztec, Inca and Mayan Myths*
(Collins & Brown, 2000)

Harper, J., *Birth of the Fifth Sun and other Meso American Tales* (Texas Tech
University Press, 2008)

Bibliography

Africa

Alcock, P.G., *Venus Rising: South African Astronomical Beliefs, Customs and Observations* (2014)

assa.saao.ac.za/wp-content/uploads/sites/23/2014/08/Venus-Rising-2014-P-G-Alcock.pdf

Bleek, W.H.I. & Lloyd, L.C., *Specimens of Bushmen Folklore* (1911)

Duane, O.B., *African Myths and Legends* (Brockhampton Press, 2004)

Hollman, J.C., '"The Sky's Things", Ixam Bushman "Astrological Mythology" as recorded in the Bleek and Lloyd Manuscripts', *African Sky*, Vol. 11 (2007) p.8

Pitcher, D., *The Calabash Child: African Folktales* (1988)

Werner, A., *Africa: Myths and Legends* (Senate, 1933)

Asia

Black, William George, 'The Hare in Folk-lore', *The Folk-Lore Journal*, Volume 1 (1883)

De Las Casas, D., *Kamishibhai Story Theatre* (Libraries Unlimited, 2006)

O'Brien, *The Jataka Tale of the Selfless Hare* (Learn Religions)

Riordan, J., *The Sun Maiden and the Crescent Moon: Siberian Folk Tales* (Interlink Books, 1989)

Shedlock, M.L., *Eastern Stories and Legends* (1920)

Quayle, E., *The Shining Princess, and other Japanese Legends* (Andersen Press Ltd, 1989)

Europe

Crossley-Holland, K., *Northern Lights: Legends, Sagas and Folktales* (Faber & Faber, 1987)

Guerber, H.A., *The Norsemen* (Senate, 1994)

Hatt, E.D. (trans. Sjoholm, B.), *By the Fire: Sami Folktales and Legends* (University of Minnesota Press, 2019)

Judd, M.C., Project Gutenberg *Classic Myths Retold* (1901)

Marshal, B.C. (trans.), *The Flower of Paradise and other Armenian Tales* (Libraries Unlimited, 2007)

Schwartz, H., *Miriam's Tambourine: Jewish Folktales from Around the World* (OUP, 1984)

Villa, S., *100 Armenian Tales and their Folk Origins* (Wayne State University, 1966)

Greek

Bullfinch, T., ed. Holme, B., *Myths of Greece and Rome* (Penguin, 1981)

Falkener, D.E., *The Mythology of the Night Sky* (Springer, 2011)

Galat, J.M., *Dot to Dot in the Sky: Stories in the Stars* (Whitecap Books, 2001)

Griffiths, A.M.M., *The Stars and Their Stories: A Book for Young People* (1913)

Kershaw, S.P., *A Brief Guide to the Greek Myths* (Robinson, 2007)

March, J., *The Penguin Book of Classical Myths* (Penguin, 2008)

Moore, P., *Legends of the Stars* (History Press, 2009)

Williams, M., *Greek Myths for Young Children* (1991)

North America

Ferguson, D., *Native American Myths* (Collins & Brown, 2001)

Harris, C., *Mouse Woman and the Mischief Makers* (Macmillan, 1977)

Hitakonanu'laxk, *The Grandfathers Speak: Native American Folk Tales of the Lenape People* (Interlink Books, 2005)

Judson, K.B. (ed.), *Myths and Legends of Alaska* (1911)

Kerven, R., *Native American Myths* (Talking Stone, 2018)

Macdonald, J., *The Arctic Sky: Inuit Astronomy, Star Lore and Legend* (Royal Ontario Museum/Nunavut Research Institute, 1998)

Millman, L., *A Kayak Full of Ghosts: Eskimo Folktales* (Interlink Books, 2004)

Monroe, J.G. & Wlliamson, R.A, *They Dance in the Sky: Native American Star Myths* (Houghton Mifflin, 1987)

San Souci, R.D., *Cut from the Same Cloth: American Women of Myth, Legend and Tall Tale* (Penguin, 1993)

Oceania

Australia

Hamacher, D.W. & Norris R.P., 'Bridging the gap through Australian cultural astronomy' in *Oxford IX International Symposium on Archaeoastronomy Proceedings* IAU Symposium No. 278 (2011) Clive L.N. Ruggles, ed.

Havecker, C., *Understanding Aboriginal Culture* (Cosmos, 1987)

McConnel, U., *Myths of the Munkan* (Melbourne University Press, 1957) (Her fieldwork in 1927 and 1934 in Gulf of Carpentaria, Cape York Peninsula, N. Queensland)

Norris, R. & Norris, C., *Emu Dreaming: an Introduction to Australian Aboriginal Astronomy* (Emu Dreaming, 2009)

Norris, R.P., 'Dawes Review 5: Australian Aboriginal Astronomy and Navigation' in *Publications of the Astronomical Society of Australia* (PASA), Vol. 33, e039, Astronomical Society of Australia (Cambridge University Press, 2016)

Parker, K.L., *Wise Women of the Dreamtime* (Inner Traditions International, 1993)

Reed, A.W., *Aboriginal Myths, Legends & Fables* (Reed New Holland, 1982)

Reed, A.W., *Aboriginal Stories* (Reed New Holland, 1994)

Smith, W.R., *Myths and Legends of the Aborigine* (Senate, 1930)

Wilson, B.K., *Tales told to Kabbarli: Aboriginal Legends Collected by Daisy Bates* (Angus & Robertson, 1972)

New Zealand

Reed, A.W., *Maori Myth and Legend* (1983)

Grace, A., *Folktales of the Maori* (Senate, 1907)

Kanawa, K.T., *Land of the Long White Cloud: Maori Myths, Tales and Legends* (Pavilion, 1989)

Pacific Islands

Flood, B., Strong, E.S. & Flood, W., *Pacific islands Legends: Tales from Micronesia, Melanesia, Polynesia, and Australia* (Bess Press, 1999)

Reed, A.W. & Hames, I., *Myths and Legends of Fiji & Rotuma* (Reed Publishing, 1967)

SOUTH AMERICA

Bierhorst, J., *The Hungry Woman* (Quill, 1984)

Boas, O.V. & Boas, C.V., *Xingu: The Indians, Their Myths* (Farrar, Straus, & Giroux, 1974)

Ferguson, D., *Tales of the Plumed Serpent* (Collins & Brown, 2000)

Harper, J., *Birth of the Fifth Sun and other MesoAmerican Tales* (Texas Tech University Press, 2008)

GENERAL

Background

Andrews, M., *The Seven Sisters of the Pleiades: Stories from Around the World* (Spinifex, 2004)

Aveni, A., *Star Stories: Constellations and People* (Yale University Press, 2019)

Brody, H., *The Other Side of Eden: Hunter-gathers, Farmers and the Shaping of the World* (Faber and Faber, 2000)

Fabian, S.M., *Patterns in the Sky: an Introduction to Ethnoastronomy* (Waveland Press, 2001)

Falkner, D.E., *The Mythology of the Night Sky* (Springer, 2011)

Frank, R.M., 'Chapter 10: Origins of the "Western" Constellations' in Ruggles, C.L.N. (ed.), *Handbook of Archaeoastronomy and Ethnoastronomy* (Springer Science and Business Media, New York)

Gooley, T., *Wild Signs and Star Paths* (Sceptre, 2018)

Krupp, Dr E.C., *Beyond the Blue Horizon: Myths and Legends of the Sun, Moon, Stars and Planets* (Harper Collins, 1991)

Simpson, P., *Guidebook to the Constellations, Telescopic Sights, Tale and Myths* (Springer, 2011)

Collections of Star/Moon/Sun Stories

Galat, J.M., *Dot to Dot in the Sky: Stories in the Stars* (Whitecap Books, 2001)

Galat, J.M., *Dot to Dot in the Sky: Stories of the Moon* (Whitecap Books, 2004)

Ganeri, A., *Star Stories: Constellation Tales from Around the World* (Templar Books, 2018)

McCaughrean, G., *The Orchard Book of Starry Tales* (Orchard Books, 1998)

Olcott, W.T., *Sun Lore of All Ages* (The Book Tree, 1914)

Ridpath, I., *Star Tales* (Lutterworth Press, 1988)

Stryer, A.S., *The Celestial River: Creation Tales of the Milky Way* (August House, 1998)

Other Collections of Stories that Contain Stars Amongst Other World Tales

Smith, Ure, *Australia's Children of the World* (1979)

Ganeri, A., *Journeys from Dream Time: Stories from the World's Religions* (Macdonald Young Books, 1998)

Useful Websites

www.ianridpath.com (Stars, their history and mythology. Reflects the book of the same name by Ian Ridpath.)

www.emudreaming.com/index.html (*Australian Aboriginal Astronomy Relates to Emu Dreaming* by Norris, R. & Norris, C., 2009)

www.wesharethesamemoon.org/ *(This project is a collaboration between storyteller Cassandra Wye and astrophysicist Dr Megan Argo to bring science into primary schools in a creative and innovative way, using stories about the moon.)*